Is: S. Brandano..

Cabo de No:

THE
PHANTOM
ATLAS

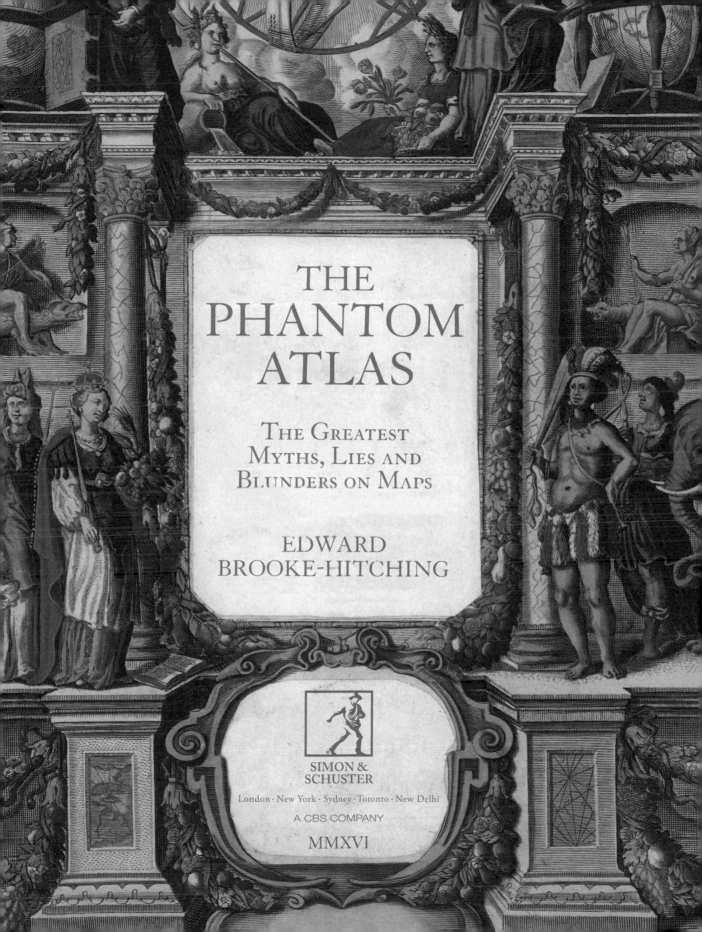

THE PHANTOM ATLAS

The Greatest Myths, Lies and Blunders on Maps

EDWARD
BROOKE-HITCHING

SIMON &
SCHUSTER

London · New York · Sydney · Toronto · New Delhi

A CBS COMPANY

MMXVI

To Emma and Franklin
Where would I be without you?

...ICAE SIVE QVARTAE ORBIS PARTIS NOVA ET EXACTISSIMA DESCRIPTIO. AVCTORE DIEGO GVTIERO PHIL ETC. COSMOGRAPHO. HIERO COCK

CONTENTS

Pages 4–5: Pieter Goos's grand Nieuwe Werelt kaert, *1672.*

Page 6: Americae Sive Qvartae Orbis Partis Nova Et Exactissima Descriptio, *by Diego Gutiérrez (1562).*

INTRODUCTION

So Geographers in Afric-maps
With Savage-Pictures fill their Gaps;
And o'er uninhabitable Downs
Place Elephants for want of Towns.

Jonathan Swift

As the sun climbed the June sky, the vessel *Justo Sierra*
cast off. Its mission: to scour the Gulf of Mexico for
the elusive 31-sq. mile (80-sq. km) island known as
'Bermeja'. The crew were following the guidance of,
among others, the cartographer Alonso de Santa Cruz,
who charted the island in his 1539 map *El Yucatán e Islas
Adyacentes*; and the more precise positioning provided by
Alonso de Chaves in 1540, in which the writer described
the land mass as 'blondish or reddish'.

Finally, they reached the given coordinates – and there they
found nothing. Only open, unbroken water, as far as the eye
could see. There was no trace of an island certified on countless
navigational charts. The mariners were thorough and swept
the area, taking extensive measurements and soundings, but
to no avail. Bermeja, it turned out, was a phantom. Just like
that, an established fact became fiction. But what is particularly
surprising about this sixteenth-century ghost territory is the
lifespan it enjoyed – because the *Justo Sierra* wasn't a ship from
antiquity – the crew was a multidisciplinary team of scientists
put together by the National Autonomous University of
Mexico. The year was 2009.

This is an atlas of the world – not as it ever existed, but as
it was thought to be. The countries, islands, cities, mountains,
rivers, continents and races collected in this book are all
entirely fictitious; and yet each was for a time – sometimes for
centuries – real. How? Because they existed on maps.

Historically, cartographic misconceptions have commonly
been disregarded. Perhaps this is because, viewed as mere errors,
there is a tendency to dismiss them as insubstantial. But one need
only glance at, say, the charts confidently proclaiming California

to be an island, the mysterious, black magnetic mountain of
Rupes Nigra at the North Pole or the depictions of Patagonia as
a region of 9ft (2.7m) giants to realize that these invented lands
are crying out for exploration. How did these ideas come about?
Why were they believed so widely? And how many other
equally strange examples are there to find?

One might assume that these ghosts have little bearing
today, but, as the story of Bermeja demonstrates, a fascinating
characteristic of many of these misbeliefs is their remarkable
durability. Indeed, there are those that survived into the
nineteenth century and beyond: Sandy Island, for example, in
the eastern Coral Sea, was first recorded by a whaling ship in 1876
and thenceforth marked on official charts for more than a century.
It finally had its nonexistence established in November 2012 –
136 years after it was first 'sighted' (and a whole seven years after
Google Maps was launched). These phantoms were considered
a plague on navigational charts, frequently leading ships astray
on fruitless confirmation missions. It was only as the ocean
highways grew busier, and global positioning more accurate,
that the methodical purging of these anomalies increased in
efficiency. In 1875, for example, no fewer than 123 nonexistent
islands (marked E.D., or 'Existence Doubtful') were cleared
from the British Royal Navy's chart of the North Pacific.

But what caused the recording of these nonexistents in
the first place? Naturally, the further back we go the more
superstitions, classical mythology and careful adherence to
religious dogma have a role to play. The complex *mappae mundi*
of Medieval Europe, for example, of which the *Hereford Mappa
Mundi* (c.1290) is the largest extant example, serve as giant
curiosity cabinets of history and popular belief. These immense,
intricate collages were for the benefit of visiting pilgrims
unable to read. Usually Jerusalem-centric, the maps were
more to illustrate the scale of God's works, with transcription
errors abounding, as well as depicting the more outrageous
phenomena reported by Pliny, such as the Sciapodes – a species
of man said to exist in the land of Taprobana, who used their
one giant foot to shade themselves from the midday sun.

Mirages and other visual phenomena have also proven
instrumental in manifesting the immaterial on maps. At sea,
formations of low clouds were mistaken for land so often that
sailors took to referring to them as 'Dutch Capes'. The Fata
Morgana in particular is a complex form of superior mirage that,
from a ship's bow, appears as a band of territory on the horizon.

The name gives some indication of how contemptuously, and fearfully, it was held by mariners: the term comes from the Italian for Morgan le Fay, the Arthurian trickster enchantress. Most often seen in polar regions, the optical illusion is a prolific culprit in the perpetration of false land sightings – it is accused, for example, of being the implement of disaster in Baron von Toll's 1902 expedition to find Sannikov Land in the Arctic Ocean.

And then, of course, there is the honest error, which is usually rooted in educated guesses of 'wishful mapping' or the limited ability of contemporary measurement systems. Coordinates were rough and imprecise, until John Harrison's invention of an accurate marine chronometer in the eighteenth century provided a long-sought solution to the problem of measuring longitude. Errors were copied, and discoveries even frequently 'remade'. Lieutenant Charles Wilkes, for example, during an 1838 survey of the Tuamotos, discovered an island at 15°44'S, 144°36'W. He named it King Island, in honour of the lookout who had spotted it. It wasn't until later that it was learnt the island had, in fact, been sighted several years earlier, in 1835, by Captain Robert Fitzroy of the *Beagle*, and named Tairaro.

Sometimes, phantoms even appear out of sheer whimsy. In his *Cosmography* (1659), Peter Heylyn tells the story of Pedro Sarmiento's capture by Sir Walter Raleigh, who subsequently interviewed the Spanish explorer about curious entries on his maps of the Strait of Magellan. Raleigh questioned his prisoner about one particular island, which seemed to offer potential tactical advantage. Sarmiento merrily replied:

that it was to be called the Painter's Wife's Island, saying that, whilst the Painter drew that Map, his Wife sitting by, desired him to put in one Countrey for her, that she in her imagination might have an island of her own. His meaning was, that there was no such Island as the Map pretended. And I fear the Painter's Wife hath many Islands and some Countreys too upon the Continent in our common Maps, which are not really to be found on the strictest search.

Also to blame are the low-down, dirty liars: those who make the calculated and committed decision to invent an entire island or country for dishonourable and self-serving purposes. The impostor George Psalmanazar, for example, was a Frenchman on a mission to hoodwink the eighteenth century. He pretended to be a resident of Formosa (Taiwan) in a deception of depth and detail that fooled many. His book, *An Historical and Geographical*

Description of Formosa, was filled almost entirely with fantastic details pulled straight from his fertile imagination.

Wild tales sold books and earned popularity. Adventurers cast themselves in heroic light, seducing funds from backers for future expeditions. Benjamin Morrell, known commonly as 'the biggest liar in the Pacific', returned from voyages breathless with tales of newly discovered lands (emblazoned with his name wherever possible) that no one else could find, with travel accounts that are clearly and liberally plagiarized. But lord of liars has to be the Scotsman Gregor MacGregor, an exaggerator and fantasist of breathtaking audacity. The corvine-eyed con-artist strode into London presenting himself as the 'Cazique of the Territory of Poyais', and proceeded to commit the greatest fraud of the nineteenth century, if not of all time.

Cartographers themselves have even indulged in minor deceptions for protection, devising their own fictitious geographies to use as copyright 'traps' in the same way as lexicographers have included fictitious entries to prove rivals have stolen their material. This isn't a solely antiquated practice, either. In 2005, a representative of the Geographers' *A–Z Street Atlas* revealed to the BBC that the London edition of their map book at that time contained more than 100 fabricated streets.

Investigating geographic ghosts can also lead to the discovery that their labelling as such can be too hastily applied: in volcanically active regions, the sudden creation and destruction of islands can be relatively common occurrences. Among cultures in these areas there are stories passed down out of oral tradition that act as records of such islands' existence: in Fiji, for example, there is the story of the inhabited island of Vuniivilevu, which one day vanished into the depths of the Pacific Ocean. To this day, when fishing boats pass over its supposed former location, the custom is to fall respectfully silent. Sometimes, the record of such disaster is a map: in the Icelandic waters there were Gunnbjörn's skerries, a group of islands home to eighteen farms that, according to a note on Johannes Ruysch's 1507 map, were 'completely burned up' by volcanic action in 1456.

However certain we are of the world around us, it seems there is always more to the story. How many other phantoms, I wonder, are hiding in plain sight, printed so assuredly on wall maps around the world? What island, what mountain, what work of imagined nation is masquerading as fact, enjoying its quiet nonexistence, just waiting to be undiscovered?

STRAIT OF ANIAN

48°29'N, 124°50'W

Also known as Strete of Anian

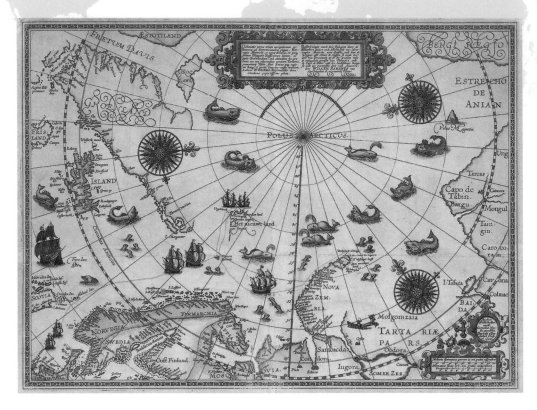

One of the greatest obsessions in the history of European exploration was the search for the Northwest Passage. Uncovering a trade route through the crushing pack ice of the Arctic to reach Asia and her endless riches – as an alternative to the gruelling and dangerous route around South America – would bring incalculable wealth to the nation that found the way. For centuries such a way was purely theoretical, willed into mythical existence through sheer mercenary desire. It wasn't until 1850 that a true Northwest Passage was discovered by Robert McClure, and until 1906 that the sea route was successfully navigated by the Norwegian explorer Roald Amundsen. But, in the centuries before this, a variety of legendary inlets and waterways potentially leading to this crossing were rumoured, depicted and pursued at great cost. The grandest of these was the Strait of Anian.

Willem Barentsz's landmark 1598 map of the Arctic region, drawn from his observations made during his 1596 voyage. It is decorated with sea monsters, ships, whales and the mythical 'Estrecho de Anian' in the top right corner.

Rumours of this strait between northwestern North America and northeastern Asia (similar to the Bering Strait) that could possibly be the western end of an Arctic passage began to appear on maps in the mid- to late fourteenth century, and inspired voyages by explorers including John Cabot, Sir Francis Drake, Gaspar Corte-Real, Jacques Cartier and Sir Humphrey Gilbert. The name 'Anian' is thought to originate from the thirteenth-century stories of Marco Polo: in Chapter 5, Book 3 of his *Travels*, the explorer mentions a gulf that 'extends to a distance of two months' navigation along its northern shore, where it bounds the southern part of the province of Manji, and from thence to where it approaches the countries of Ania, Tolman and many others already mentioned'. He describes its geography in detail, before concluding: 'This gulf is so extensive and the inhabitants so numerous, that it appears like another world.'

Here Polo is referring to the Gulf of Tonkin, off the coast of northern Vietnam, and, although clearly suggesting it to be located a good deal farther south, it is easy to understand how cartographers searching for information on the area grabbed the name 'Ania' to fit reports of a strait in the general vicinity. It first appeared in a work by the Italian cosmographer

The earliest printed map to focus solely on North America, and the first to show the Strait of Anian (Streto de Anian), separating America and Asia. It was by Paolo Forlani and Bolognino Zaltieri, Venice (1566).

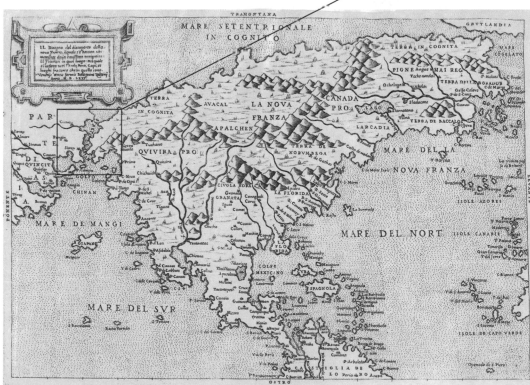

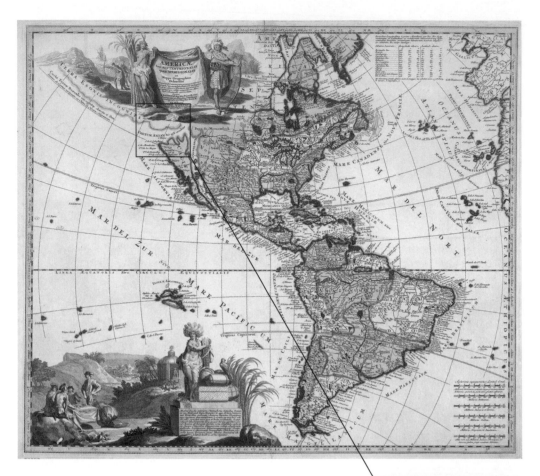

Giacomo Gastaldi in 1562, and was then adopted by the mapmakers Bolognini Zaltieri and Gerardus Mercator in 1567. The dream of the Strait of Anian was held onto tightly by explorers and cartographers over the next few hundred years, because of its theoretical instrumentality in finding the elusive Northwest Passage. European trade with Asia was booming but it was a demanding task, for goods had to be carried over land or sailed around the Cape of Good Hope. The latter, an especially terrible risk to shipping, was originally named 'Cabo das Tormentas' ('Cape of Storms') by the Portuguese explorer Bartolomeu Dias in 1488.

The Greek seaman Juan de Fuca (1536–1602) was one of several men who claimed to have sailed the Strait of Anian. Under the orders of the viceroy of New Spain, de Fuca launched two expeditions to find the fabled way. The first, consisting of three ships carrying 200 men, is recorded as failing in the early stages when the crew took the ship to California after a mutiny over the captain's 'malfeasance'.

Adam Zuerner's Americae tam Septentrionalis quam Meridionalis in Mappa Geographica Delineatio *(c.1707), with the 'Fretum Anian' drawn just below the cartouches of the Native American hunters.*

A second attempt was made in 1592, when the viceroy ordered de Fuca to return to the region with two ships; it was supposedly more successful. According to the merchant Michael Lok, de Fuca:

came to the Latitude of fortie seven degrees, and that there finding that the land trended North and north-east with a broad inlet of sea, between 47 and 48 degrees of Latitude; he entered thereinto, sayling therein more than twenty days, and found … very much broader Sea than was at the said entrance, and that he passed by divers lands in that sayling …

De Fuca recorded the opening of the strait as guarded by a large island with a towering rock spire; he then returned jubilant to Acapulco in the hope of gaining a reward for his findings, but none was offered.

Decorative example of Ortelius's map of the Tartar kingdom in 1598, with the 'Stretto di Anian' drawn just east of centre.

*Cornelis de Jode's 1593
depiction of the west coast
of North America.*

Because the sole written source for de Fuca's travels is that of Lok, an Englishman who claimed to have met the sailor in Venice (and who was a keen promoter of the search for the passage), there is some doubt as to whether de Fuca ever actually existed – some scholars have deemed him as legendary as his findings. And yet, if he was fictitious, there are curiously accurate elements to his geography. In 1787, a fur trader named Charles William Barkley discovered a strait on the west coast of North America at Cape Flattery and, although a full degree (roughly 69 miles/111km) farther north than de Fuca had claimed, he recognized it as the waterway de Fuca had reported by spotting the pinnacle the sailor had described (which is now known as the De Fuca Pillar). De Fuca's alleged discovery of the Anian Strait was backed up by the Spanish navigator Lorenzo Ferrer Maldonado, who claimed to have sailed the waterway in the opposite direction in 1588, four years before de Fuca. (Although Maldonado's account is clearly fabricated, and achieved little recognition at the time, its rediscovery in the late eighteenth century gave the strait renewed fame.) This waterway that Barkley discovered was named the Strait of Juan de Fuca, but it was merely a 95-mile (153km) long passage that functions both as the Salish Sea's outlet to the Pacific and as the starting point of the international boundary between America and Canada.

The desperate hunt for a transcontinental passage meant that the Strait of Anian haunted maps for hundreds of years. A 1719 map by Herman Moll suggests it as a bay 50° north of the Island of California (see relevant entry on page 64). The 1728 edition of a map by Johannes van Keulen also places it here, accompanied by the note: 'They say that one can come through this strait to Hudson Bay, but this is not proven.' In 1772, Samuel Hearne travelled over land from Hudson Bay to Copermine River and back – an extraordinary voyage of more than 3600 miles (5800km) – in search of the channel, but no Strait of Anian was found. For all but the most hopeful, this was sufficient to lay the myth to rest.

ANTILLIA

33°44'N, 54°55'W   *Also known as Antilia, Isle of Seven Cities, Ilha das Sete Cidades, Sept Citez*

In 711, the Islamic Moors of north Africa crossed the Strait of Gibraltar and invaded the Iberian peninsula. Led by the general Tariq ibn Ziyad, this massive force waged an eight-year campaign, crushing the Visigothic Christian armies and bringing most of modern Spain and Portugal under Islamic rule. The Moors continued their rampage across the Pyrenees, eventually falling to the Franks led by Charles Martel at the Battle of Poitiers in 732; but before that a strange legend emerged from the rubble of their Spanish invasion. It told of a group of seven Christian bishops who managed to flee the Muslim forces by ship across the Atlantic, eventually taking refuge on a distant island known as 'Antillia'. There, the holy men decided to set up residence, and each built for himself a magnificent golden city. This gave the island its other name: 'Isle of Seven Cities'.

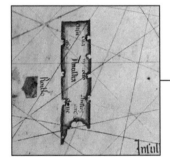

How the bishops fared on the island is unknown, for no mention of Antillia is made for another seven centuries, until it began to appear on maps such as the *c.*1424 portolan (sailing instructions) chart of the Venetian cartographer Pizzigano, which shows several of these legendary Atlantic Islands. Here, Antillia is depicted as a large, rectangular block, with seven cities adorning its coasts: Asay, Ary, Vra, Jaysos, Marnlio, Ansuly and Cyodne. Supposedly, the vast island was located in the North Atlantic, 750 miles (1400km) west of Portugal in the latitude of Gibraltar. The origin of its name is equally mysterious, but is thought to derive from *anteilha*, roughly 'opposite isle', possibly because it was thought to lie across from the Portuguese coast. (The name would later be applied to the Antilles Islands.)

The considerable size of the island made it attractive to explorers: Portugal's Prince Henry (1394–1460), better known as Henry the Navigator, dispatched a captain named Diogo de Teive and the Spanish nobleman Pedro de Velasco in 1452 to sail from the island of Fayal in the Azores, and make southwesterly and northwesterly sweeps in search of Antillia. The men reached as far as the latitude of southern Ireland,

without so much of a glimpse of the Antillian shore. The mission wasn't a total loss, however: during their journey they discovered Corvo and Flores, two outer islands of the Azores. In a letter to Ferman Martins in 1474, the Italian astronomer Paulo Toscanelli stated his certainty that the island of Antillia could be found 50 degrees east of Cipangu (Japan), and recommended it as a convenient waypoint when journeying to Cathay (China). Then, in 1486, King João II gave permission to Fernão Dulmo, captain of the northern territory of Terceira (one of the larger islands of the Azores), to locate and claim the Isle of Seven Cities in his name. Dulmo launched a search party in March but found nothing but terrific storms.

The portolan chart by Albino de Canepa (1489), with Antillia featured as a rectangular island to the far left.

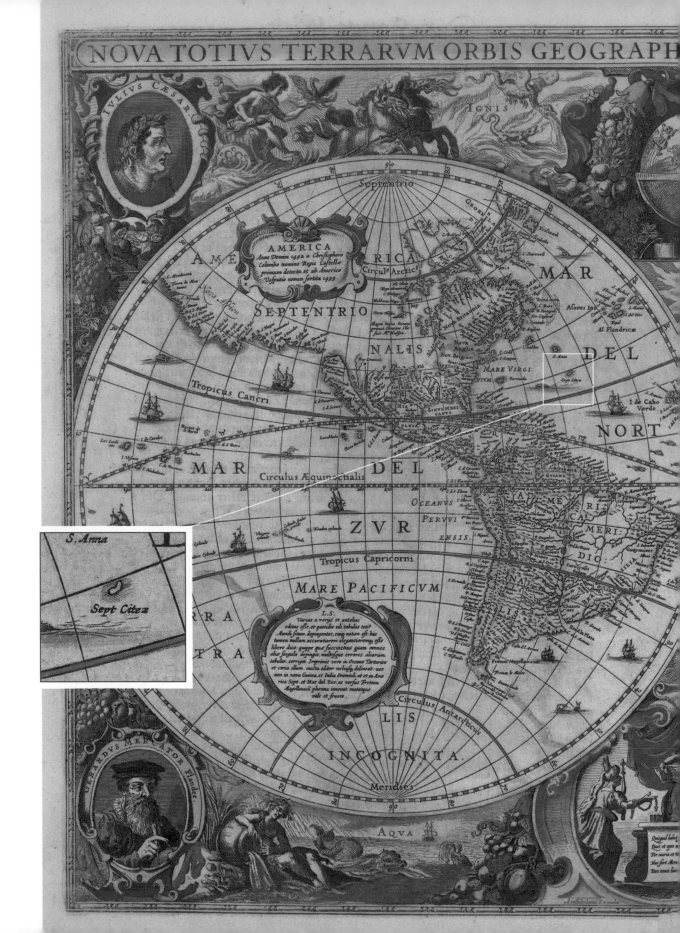

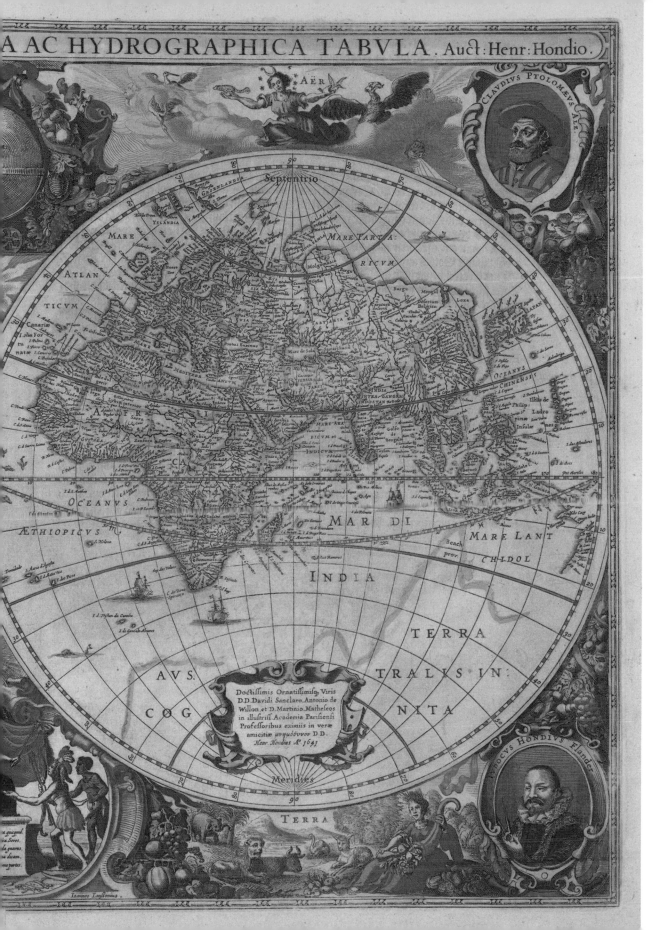

A AC HYDROGRAPHICA TABVLA. Auct: Henr: Hondio.

Doctissimis Ornatissimisq; Viris
D.D. Davidi Sanclaro, Antonio de
Willon, et D. Martinio, Matheseos
in illustriss. Academia Parisiensi
Professoribus eximiis in veræ
amicitiæ μνημόσυνον D.D.
Henr. Hondius Aᵒ. 1641

Columbus, too, believed in the existence of Antillia, and reckoned the island a useful stopping-off point en route to the Indies. Entries in his travel journal of 1492 suggest he expected to find it at 28°N. This would have been based on the position given by Martin Behaim, who that same year had made the first cartographic mention of the island on his 'Erdapfel' (literally 'earth-apple') globe, with the note:

In the year 734 after the birth of Christ, when all Spain was overrun by the African heathens, this island of Antillia, called also the Isle of Seven Cities, was peopled by the Archbishop of Porto with six other bishops, and certain companions, male and female, who fled from Spain with their cattle, property and goods. In the year 1414 a Spanish ship approached very near this island without danger.

In 1508, additional Antillian detail was supplied by Johannes Ruysch on his map, with the inscription:

This island Antilia was once found by the Portuguese, but now when it is searched, cannot be found. People found here speak the Hispanic language, and are believed to have fled here in face of a barbarian invasion of Hispania, in the time of King Roderic, the last to govern Hispania in the era of the Goths. There is one archbishop here and six other bishops, each of whom has his own city; and so it is called the island of seven cities. The people live here in the most Christian manner, replete with all the riches of this century.

Hernando Colón, son of Columbus, was also fascinated by Antillia. He suggests convincingly in his *Historia del Almirante* (1571) that the exodus of the bishops took place in 714, not 734, which would line up closer with the two-year rule of King Roderic in 711. He also writes that the holy men burnt their ships on arrival at Antillia, lest they should ever consider returning to Hispania. So how did the story of the exiled bishops find its way to the mainland? Colón relates the story that, during the rule of Prince Henry, a wayward ship blown off course by a storm landed at Antillia. The crew explored the island, greeted the locals and attended a church service before hurrying back to Portugal to report the experience. However, when they were ordered to return to the island for confirmation, the entire crew disappeared. Elsewhere, the French sailor Eustache de la Fosse heightened the mystery by warning that Antillia was protected by a spell placed by one of the bishops 'knowing the art of necromancy', and predicted that the island would not be found again until 'all Spain should be restored to our good Catholic faith'. De la Fosse also claimed that sailors passing the invisible island had reported shore birds flying over their vessels, although these were also invisible 'because of the said enchantment'.

Antillia next crops up in Antonio Galvão's *The Discoveries of the World* (1563), in which the chronicler shares an account of a Portuguese ship from the Strait of Gibraltar encountering an island of seven cities. The inhabitants, who spoke Portuguese as their native tongue, enquired as to whether Spain was still ruled by the Moors, from whom they had fled after the death of King Roderic. Upon returning to Lisbon, the ship's captain gave a sample of the island's soil to a goldsmith for analysis, who declared the earth to be composed of two parts soil, one part gold. (This last detail, however, is a common addition to expeditionary stories to stir up interest, and it is clear from Galvão's distanced tone that he was wary of the tale). To Galvão, it was evident that Antillia had been confused by sailors with the Caribbean Antilles far to the west. This conclusion was supported by other geographers of the period, and the island began to be cleared from maps, although one finds it occasionally included on later works, such as Hondius's stunning 1631 map of the world.

A map of Atlantis by Bory de St-Vincent, taken from Sur les Canaries *(1803).*

All missing islands of the past pale in scale to the largest and most famous fugitive of all: the island of Atlantis – 'larger than Libya and Asia put together', according to Plato, whose two dialogues, *Timaeus* and *Critias* describe the land in detail. Written by the philosopher *c*.360 BC, the works serve as the earliest record of the tale – 'not a fiction but a true story' – in which is discussed a massive war between the ancient Athenians and the Atlanteans waged 9000 years before Plato's time. Plato

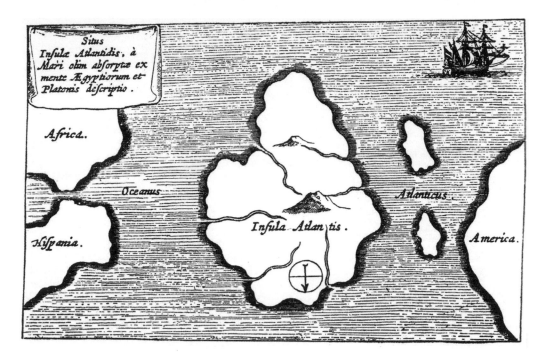

A map by the seventeenth-century scholar Athanasius Kircher, placing Atlantis equidistant between Africa and America.

uses the story of Atlantis as an allegory for the arrogance of powerful nations, drawing inspiration, it is thought, from the volcanic destruction of the island of Thera (Santorini) that occurred in the mid-second millennium BC. Aristotle dismissed it as fiction, but the Greek academician Crantor ardently defended it as historical truth. Debate then raged (and, in some quarters, still does) as to whether the tale has factual basis.

In *Timaeus*, Plato writes of a mighty island power 'situated in front of the straits which are by you called the Pillars of Heracles' (the entrance to the Strait of Gibraltar) that launched an unprovoked attack on the whole of Europe and Asia. In response, the state of ancient Athenians:

shone forth, in the excellence of her virtue and strength, among all mankind. She was pre-eminent in courage and military skill …
she defeated and triumphed over the invaders, and preserved from slavery those who were not yet subjugated, and generously liberated all the rest of us who dwell within the pillars. But afterwards there occurred violent earthquakes and floods; and in a single day and night of misfortune all your warlike men in a body sank into the earth, and the island of Atlantis in like manner disappeared in the depths of the sea. For which reason the sea in those parts is impassable and impenetrable, because there is a shoal of mud in the way; and this was caused by the subsidence of the island.

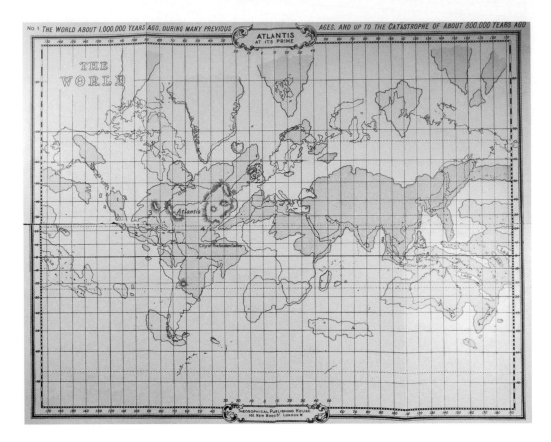

Atlantis in its prime, *from W. Scott-Elliot's* The Story of Atlantis and the Lost Lemuria *(1925).*

The allegory of the superiority of Plato's ideal state was lost among the assurances of veracity made by its author, and the intoxicating excitement that ensued. Atlantis slowly came to represent all the lost worlds and utopias ever rumoured, amalgamated with myths across cultures. 'It was a legend so adapted to the human mind that it made a habitation for itself in any country,' wrote Dr Jowett, the renowned nineteenth-century translator of Plato. 'It was an island in the clouds, which might be seen anywhere by the eye of faith … No one knew better than Plato how to invent a noble lie.' As such, there have been a plethora of academic (and less-than-academic) theories offered as to the real location of the disappeared race, including Peru, the West Indies, Antarctica, the Canary Isles, Cuba, Indonesia, Nigeria, Morocco, Cyprus, Sri Lanka, Sardinia, North America and the English Channel.

The maps here are rare instances of the legend committed to cartography – Athanasius Kircher follows Plato's description to depict it in the centre of the Atlantic Ocean. The German scholar included it in his extraordinary book *Mundus Subterraneus* (1665), which also features other mythical identifications: such

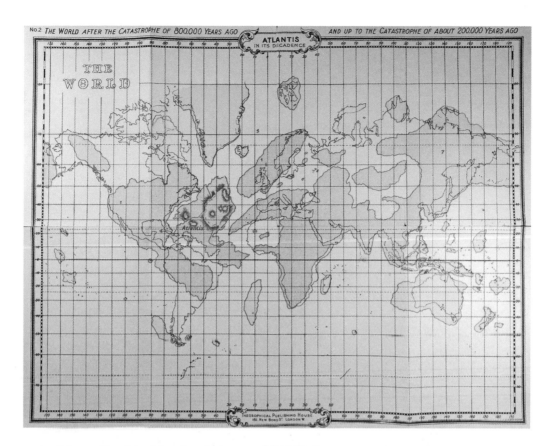

as the 'Mountains of the Moon' as the source of the Nile (see Mountains of the Moon entry on page 162); discussions on the buried remains of giants; and a commentary on the creatures of the underground world, including dragons. It is a work perhaps most famously known for the illustration *Systema Ideale Pyrophylaciorum,* a study of the Earth's volcanic system, a planet 'not solid but everywhere gaping, and hollowed with empty rooms and spaces, and hidden burrows', with terrible volcanoes being 'nothing but the vent-holes, or breath-pipes of Nature'.

The myth of Atlantis endured, albeit pushed past the border of the scholarly into the realm of the obsessed and eccentric. In *Reflections of a Marine Venus* (1953), Lawrence Durrell writes about discovering a list of diseases as yet unclassified by medical science, 'and among these there occurred the word islomania, which was described as a rare but by no means unknown affliction of spirit. There are people ... who find islands somehow irresistible. The mere knowledge that they are on an island, a little world surrounded by the sea, fills them with an indescribable intoxication. These born "islomanes" ... are direct descendants of the Atlanteans.'

And here Scott-Elliot maps Atlantis in its 'decadence'.

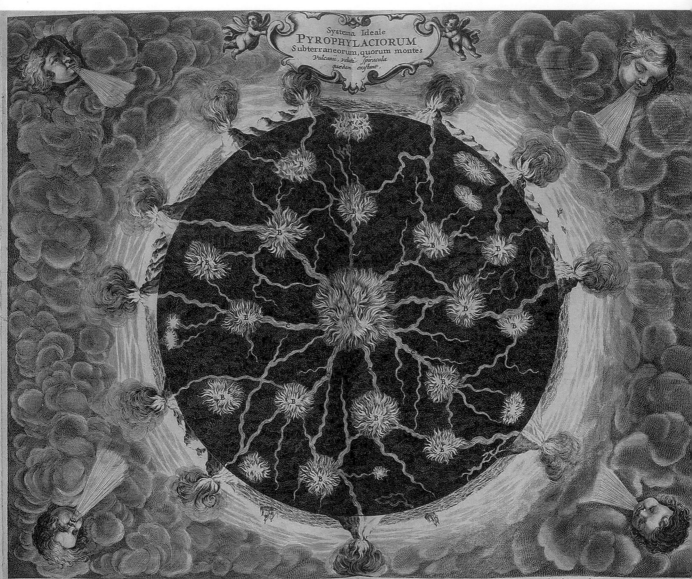

Athanasius Kircher's 1665
depiction of the 'fire canals',
or volcanic system, of the
subterranean world.

There has certainly been a specific islomania pertaining to those seeking Atlantis, which took an especially strange form in the creation of the 'Principality of Atlantis' by a group of Danish Atlantomanes led by John L. Mott in 1917. To escape war-torn Europe, the men claimed they had settled on a group of islands 200 miles (370km) southwest of Florida, eight degrees north of the Equator and 3 miles (5.5km) offshore of Panama and Costa Rica, which they declared their 'private Dynasty … or Principality of Atlantis Kaj Lemuria'. These details come to us from a US government file containing two decades of correspondence between the US State Department and various persons on the subject of the Atlantis principality between the 1930s and 1950s. This includes one document bearing the letterhead 'Government of Atlantis and Lemuria', in which the governor-general of the principality, a Miss Gertrude Norris Meeker, warns the US State Department that 'any trespassing in these islands or Island Empire is a prison offence'; while another letter in the file, from 1957, advises the government to respect the sovereignty of the principality: 'Believe me,' writes Leslie Gordon Bell, legal counsel of the new Atlanteans, 'this is not a figment of somebody's imagination.'

AURORA ISLANDS

52°37's, 47°49'w

In 1762, the Spanish merchant ship *Aurora*, captained by
José de la Llana, was on its way home to Cádiz from a
mission to Lima when the crew sighted a pair of islands
midway between the Falkland Islands and South Georgia.
Traffic was increasing in this region due to its proximity
to the route taken by European trade vessels to round
Cape Horn and so, unsurprisingly, the Auroras were
confirmed by a succession of crews on trading missions:
the frigate *San Miguel* spotted them in 1769, as did the
Aurora again in 1774, followed by the *Perla* in 1779 and
the *Dolores* in 1790, marking the coordinates using dead
reckoning, which is essentially skilful guesswork.

In 1790, the *Princessa* of the Royal Philippine Company,
ferrying goods from Spain, also reported passing the islands on
its voyage to Lima – Captain Manuel de Oyarvido provided
precise coordinates, and recorded the existence of a third,
which he named 'Isla Nueva'. The Spanish explorer José
Bustamente Guerra was then instructed to chart the 'islas
Aurora', and in 1794 he found an island at 52°37'S with a
snow-covered eastern side, and a western side dark with snow
spreading through its ravines. His ship, the *Atrevida*, cruised
alongside the island making observations from only 1 mile
(1.8km) offshore, before continuing on its way. Four days later,
he came across a second island, and from 'a moderate distance'
noted a snow-covered southeast side. Satisfied that the islands
were now placed with geographic precision, Bustamente
proceeded to Montevideo. His charts were handed to the Royal
Hydrographical Society of Madrid, where they were studied
and filed away; and for a time no more was thought of them.

   Over twenty years later, in 1820, the British sailor and seal-
hunter James Weddell was drawn to the area by Bustamente's
survey. He arrived at the coordinates and found open water.
Refusing to believe that so many sailors before him could have
made such a major error, he patrolled the area for four days
until deciding that 'the discoverers must have been misled by
appearances', and continued on around the Falkland Islands.

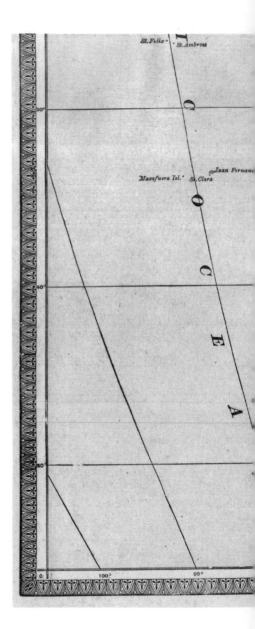

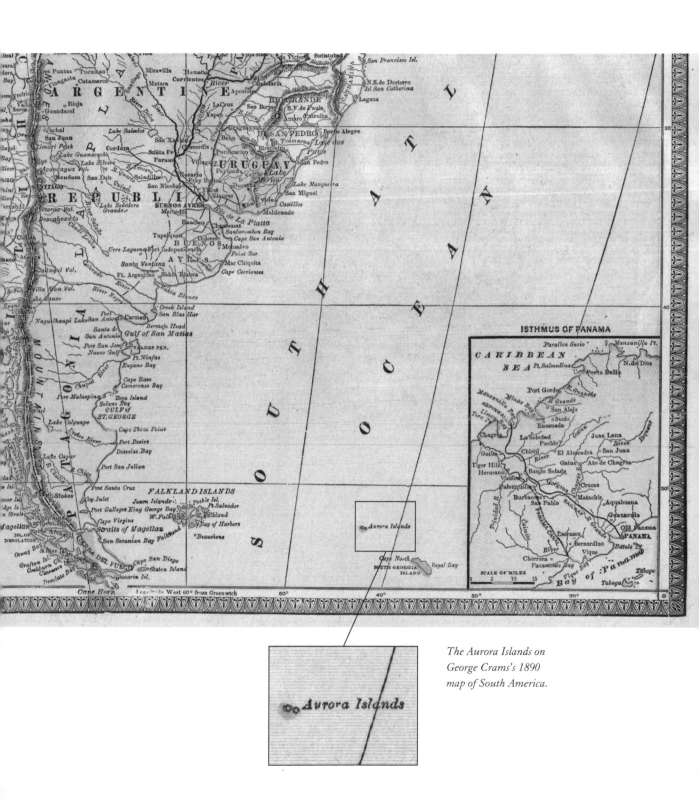

The Aurora Islands on George Crams's 1890 map of South America.

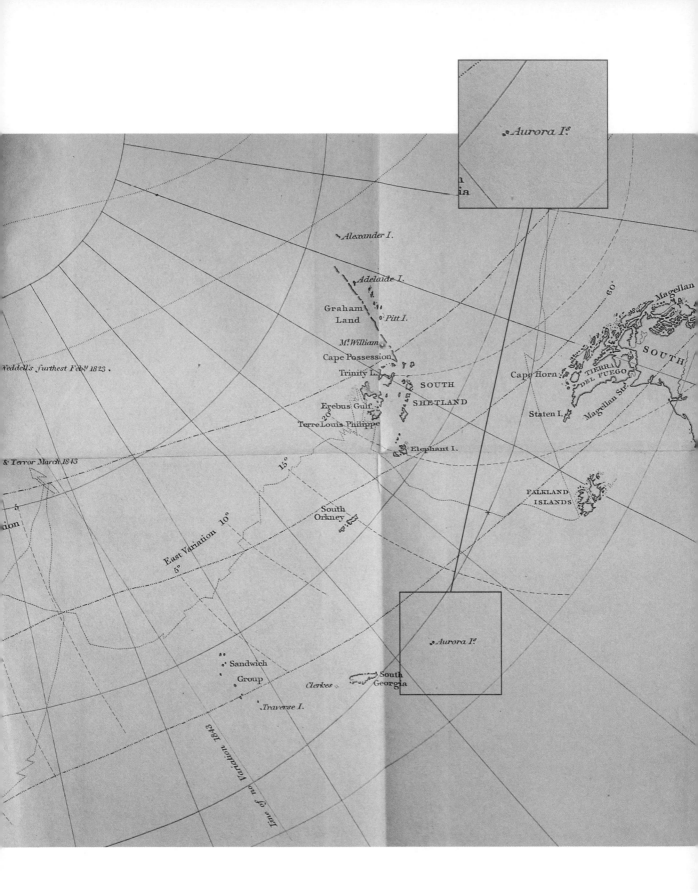

Aurora I.

ia

Alexander I.

Magellan

60

Adelaide I.

Graham
Land   Pitt I.

SOUTH

Mt. William
Cape Possession

Cape Horn   TIERRA
DEL FUEGO

Trinity I.

Weddell's furthest Feby 1823.

SOUTH
SHETLAND

Staten I.   Magellan Str.

Erebus Gulf
20°
Terre Louis Philippe

Elephant I.

& Terror March 1843

FALKLAND
ISLANDS

15°

South
Orkney

East Variation 10°

6°

Aurora I.

Sandwich
Group

Clerkes   South
Georgia

Traverse I.

Line of non Variation 1843

Weddell was right – not only are there no islands in this general vicinity, no satisfactory explanation for how the Auroras came into reputed existence has ever been provided. There are plenty of factors to take into consideration: low visibility in difficult Antarctic weather conditions; the desperation to find land on the horizon during long periods at sea; perhaps even destruction by volcanic action. Were the islands in fact giant floating icebergs, or 'ice-islands incorporated with earth' as Weddell eventually concluded? Or were they confused with the sealer's other discovery 620 miles/1150km off the Falkland Islands at 53°33'S, 42°02'W – the Shag Rocks (which were also then given the Spanish name Islas Aurora)? It has also been suggested that the Auroras might have been confused with the Falklands, but for so many skilful mariners to make the same grievous blunder seems most unlikely. The islands pop up on maps into the nineteenth century, inspiring further futile searches by Benjamin Morrell in 1823, and by John Biscoe in 1830. They can also be found on the chart accompanying John Ross's 1847 *A Voyage of Discovery and Research to Southern and Antarctic Regions.* By 1856, they had been wiped from official cartographic records. The mystery as to what so many men saw on those waters has never been solved.

Opposite: Chart from John Ross's A Voyage of Discovery and Research to Southern and Antarctic Regions *(1847).*

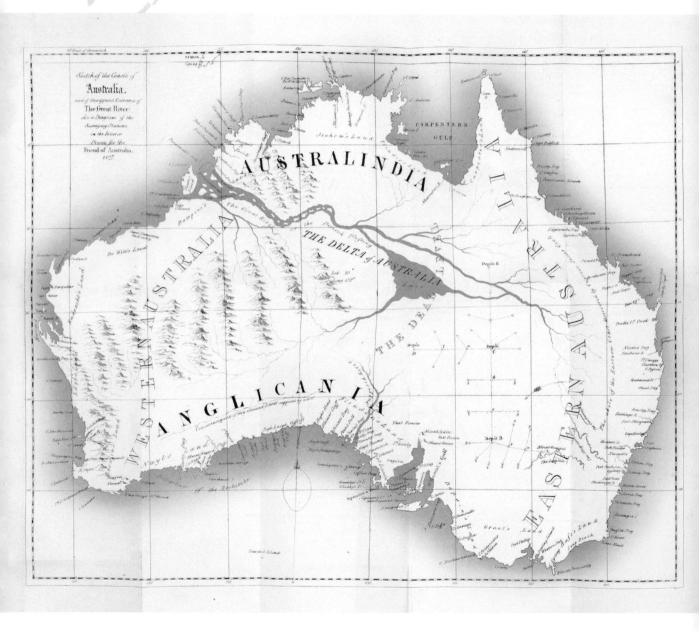

Maslen's wishful mapping of Australia's possible inland sea and river system, from The Friend of Australia *(1830).*

It had been forty-two years since the British First Fleet, commanded by Captain Arthur Philip, landed at Australia's Botany Bay and formed the first European colony at Port Jackson. Initially, the new land served as a penal territory, but the British were keen to push deeper into the unmapped Australian interior and get a sense of the potential for further settlement. They knew from experience that following rivers inland usually led to mountains, river systems and fertile land that frequently exceeded expectation, and so it was assumed that the same topographic logic could be applied to Australia – what kind of rich verdant paradise, it was wondered, could be waiting in the heartland?

Maslen's scene of colonists crossing Australia's possibly bountiful desert river system, with horses transporting tub-like vessels.

Drawn for the Friend of Australia.   Printed by C.Hullmandel.

THE EXPEDITION CROSSING A RIVER IN AUSTRALIA.

'The plan here offered is a practical scheme', announces the English writer Thomas J. Maslen in *The Friend of Australia* (1830), 'and not a vain theory which could not be put into practice; and it will serve equally well as a guide and book of reference, to a numerous or a small party of explorers.' Maslen, a retired officer of the East India Company, wrote his book to encourage colonial expansion efforts. It provides detailed instructions for how to conduct surveys and inland exploration (for the latter, he recommended the use of camels). It seemed most unlikely to Europeans that a country the size of Australia would exist without the same abundant water systems as that of other continents. Maslen, therefore, used his book to exhibit his educated guess of a water-rich Australian interior. Today, *The Friend of Australia* is considered the ultimate monument to speculative geography.

The map shown on page 34 is the one that accompanied the book, and which contributes to the rare work's modern reputation as a curiosity (only 250 copies were printed, and even those failed to sell out at the time). The theory of a vast, undiscovered Australian system of rivers and lakes had been popular for years, but it was Maslen who ran away with the idea in spectacular fashion. In the appendix, he describes the thinking behind the

Maslen's flag design 'Respectfully submitted for the consideration of Government for the adoption of the Colony of New South Wales'.

How an expedition with camels through Australia's deserts might look.

THE EXPEDITION in a DESERT in AUSTRALIA.

creation of his ideal 'atlas of Australasia *a desideratum*', supposing there to be a succession of hills stretching from the west coast towards the interior. These perhaps enclosed a high table of land, 'from whence other streams might direct their course to the dead level, and perhaps form one or more sheets of water, as the formation of lakes is one of nature's great features in Australasia'. The river network portrayed is a wonderfully elaborate and generous fantasy, crowned by a great lake the size of a small sea that is placed plum in the desolate centre of what is now known as the Simpson Desert.

Though the hydrographic ambition seemed to ignore everything known about Australia's aridity, explorers were, nonetheless, inspired to investigate. Charles Sturt was one such water-hunter, who led expeditions in 1829–30, certain that the western-flowing waterways would lead him to a giant inland sea, not unlike the large 'Delta of Australia' drawn by Maslen – but he returned disappointed. Sturt, eventually, solved the mystery with the discovery that the western channels were, in fact, tributaries to the Murray, Australia's longest river. By the middle of the nineteenth century, the inland sea myth had finally dried up.

Transporting canoes across the countryside, for use on the theoretical inland sea.

Drawn for the Friend of Australia.   Printed by C Hullmandel
CARRYING LARGE CANOES with the EXPEDITION in AUSTRALIA .

BERMEJA

22°33'N, 91°22'W

Also known as Vermeja

There is a curious phenomenon in marine law known as a 'Donut Hole'. Donut Holes are legal loopholes created by the passing of a 1982 UN convention on the Law of the Sea, which essentially states that the area of water within 200 miles (370km) from the coast of a country is the Exclusive Economic Zone (EEZ), or nautical sovereignty, of the respective nation. The Donut Holes appear when the perimeters of the EEZs of two countries don't quite meet – accordingly, these no-man's lands are deemed pockets of international water.

In the Gulf of Mexico, several of these Donut Holes, or 'Hoyas de Donas', were formed, and quickly became a point of contention between the United States and Mexico for one reason: oil. The gulf's oil fields are especially bountiful, and crucial to both countries – a current fact sheet from the US Energy Information Administration records the area as providing 17 per cent of America's total crude oil production. In this rush to clarify rights to the fields, antique maps of the region were suddenly called upon to play a crucial role in an international debate that would result in substantial wealth for the victor. Since the sixteenth century, it was discovered, charts showed a small Mexican island named 'Bermeja' nestled deep in the gulf – its existence, though, had never been proven. The Mexicans realized that, if the island could be found, it would dramatically extend their EEZ, and qualify their claim to oil rights in the region.

Bermeja first appeared on Alonso de Santa Cruz's 1539 map *El Yucatán e Islas Adyacentes*, and through to the nineteenth-century maps of the Gulf of Mexico insisted that the island could be found off the north coast of the Yucatán peninsula. Alonso de Chaves was the first to record an exact location in his *Espejo de navegantes* (Seville, *c*.1540), describing the island from a distance as seeming 'blondish or reddish'. No confirmed sighting was reported after that, but it remained on charts into the nineteenth century, when several British maps recorded the island as having sunk mysteriously. Its last appearance is found in the 1921 edition of the *Geographic Atlas of the Mexican Republic*.

Opposite: A Map of the United States of Mexico *(1826) by* Henry S. Tanner, showing Bermeja floating in the centre of the Gulf of Mexico.

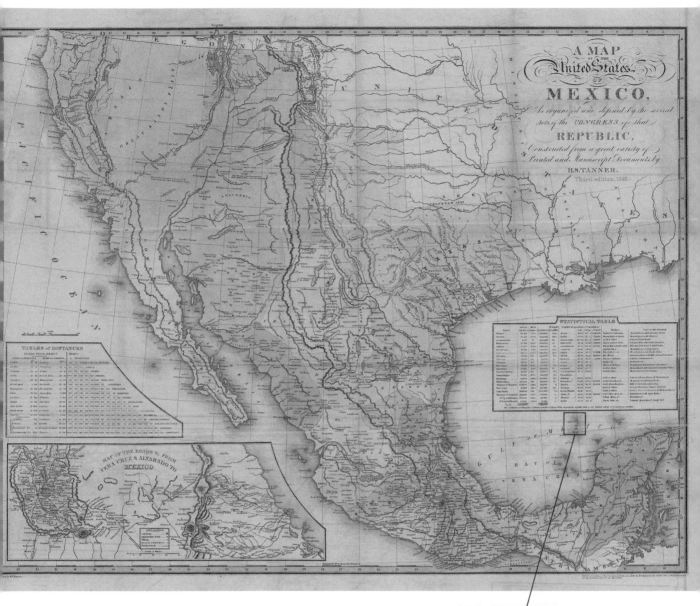

In 1997, as the United States and Mexico prepared to negotiate a treaty to divide the 'Hoyas de Donas' region, a Mexican Navy vessel was sent on a discovery mission, but was unable to find any sign of it after surveying the Yucatán waters. Mexico went on to sign the treaty in 2000, but the Mexican government never lost hope that Bermeja would one day be found, and so in 2009 a team of experts from the National Autonomous University of Mexico (UNAM) left the Mexican coast on the research vessel *Justo Sierra* to search the Mexican gulf for the 31-sq. mile (80-sq. km) island. The UNAM team arrived at the coordinates and studied the area with extended

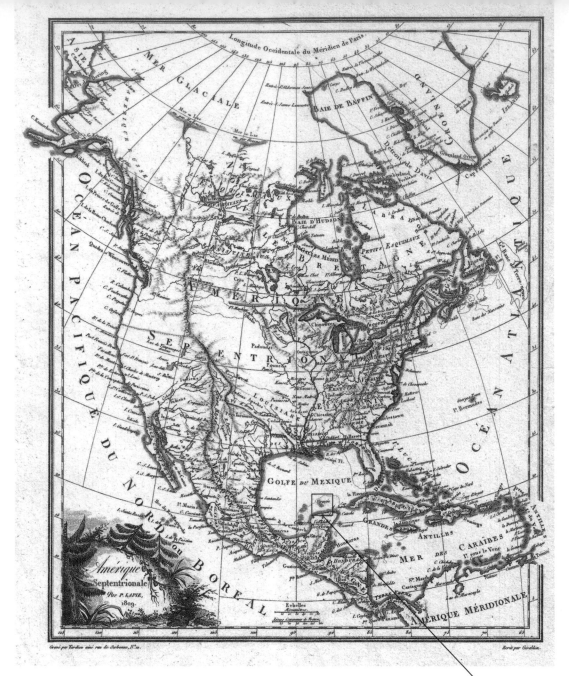

sweeps; they even deployed personnel to scan the area from above with aircraft. They found nothing but sediment-covered ocean floor.

Various theories have been suggested as to Bermeja's 'disappearance'. Some blame climate change and rising sea levels, others an undersea earthquake, although, in 2010, a group of Mexican senators released a statement pointing out that such 'a force of nature does not take place without anyone noticing, and much less so when it is sitting in an area with more than 22 billion barrels of oil reserves'.

Another popular theory is that the entire island was destroyed by the US Central Intelligence Agency to ensure US hegemony over the oil fields. In November 2000, six senators from Mexico's governing party of Partido Accion Nacional (PAN) voiced 'plentiful suspicions' on the senate floor that the island may have been purposefully vanished. The conspiracy theories flared stronger than ever in 1998 when one of the politicians, PAN party chairman José Angel Conchello, was driven off the road and killed by an assailant who was never caught, shortly after demanding further investigation into the possibility of Bermeja's existence. Conchello had warned about a secret plan by the Zedillo government to give up exploration rights to US companies.

So what has been concluded? Jaime Urrutia of UNAM and Saul Millan of the Instituto Politecnico Nacional decided that to obliterate an island of Bermeja's size a hydrogen bomb would be required. Millan suggested that, rather than destroy the island, it could have been hidden under the water, pointing to theories that the US government somehow managed to discreetly shave it down to below sea level.

Irasema Alcántara, a geographer at UNAM, passionately defended Bermeja's existence, telling reporters: 'We've encountered documents containing very precise descriptions of Bermeja's existence … On this basis we firmly believe that the island did exist, but in another location.' Julio Zamora, president of the Mexican Society of Geography, disagreed: 'Countries making maps in the 16th and 17th centuries published them with inaccuracies to prevent their enemies from using them.' This fits with the scientific opinion offered by German oceanographer Hans-Werner Schenke, of the Alfred Wegener Institute in Bremerhaven. After the UNAM team returned empty-handed in 2009, Schenke was consulted by a *Der Spiegel* journalist and delivered the final blow to Mexican hopes, pointing out: 'If you look at the latest marine maps and data of the earth, there is no indication that there ever has been an island.'

Opposite: 'Vermeja' on Tardieu's Amerique Septentrionale *(1809).*

BRADLEY LAND

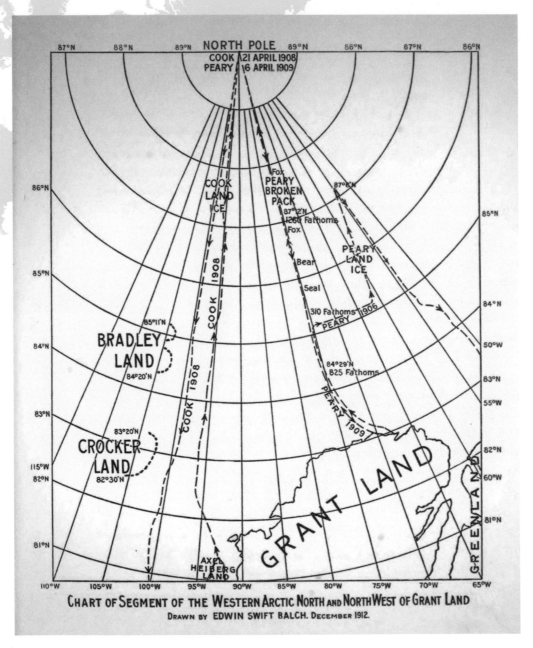

CHART OF SEGMENT OF THE WESTERN ARCTIC NORTH AND NORTH WEST OF GRANT LAND
DRAWN BY EDWIN SWIFT BALCH. DECEMBER 1912.

In the spring of 1908, the American surgeon and explorer Frederick Albert Cook left the Greenland box-house he had built upon arriving in the country the summer before, and embarked on a mission to become the first to

Chart of Segment of the Western Arctic North and Northwest of Grant Land *from Edwin Balch's* The North Pole and Bradley Land *(1913).*

reach the North Pole. With him as he crossed Smith Sound to Ellesmere Island, an Arctic archipelago north of Canada, were ten Inuit helpers, eleven sledges and 105 dogs. The convoy headed for the Bache peninsula, then followed a fjord to the west. After traversing the frozen Bay fjord they finally reached Cape Thomas Hubbard, located at the northernmost tip of Axel Heiberg Island, another in the Canadian Arctic Archipelago. (It was here at Cape Thomas Hubbard two years earlier that Robert Erwin Peary, his former friend and great rival, glimpsed the imaginary Crocker Land; see Crocker Land entry on page 70.) Cook led his group out through the slicing winds and they trudged across the frozen polar sea. Within three days, only two of his Inuit companions, Ahwelaw and Etukishook, remained by his side. They then set off for the North Pole … and all three men vanished.

Frederick Cook posing in front of an Arctic backdrop for a publicity photograph, c.1911.

For a year nothing was heard, and it was assumed the journey had met with disaster until, in April 1909, Cook suddenly reappeared. Upon his return to Anoritoq, Greenland, he told his story. He claimed to have reached Devon Island in the archipelago (and the largest unihabited island on Earth) by sledging between the islands Ellef Ringnes and Amund Ringnes. He pushed on, passing the coordinates of Peary's Crocker Land – the existence of which he refuted – and with great excitement sighted a new land mass, which he called 'Bradley Land'. The naming was in tribute to John R. Bradley, the wealthy big-game hunter who had funded Cook's expedition. Bradley Land was a large formation, said Cook, two large masses with a break, a strait or an indentation, between them. In the record of the adventure he was later to publish, *My Attainment of the Pole: Being the Record of the Expedition that First Reached the Boreal Center, 1907–1909*,

Portrait of Robert Peary, 1909.

Cook included two photographs of Bradley Land, with the description: 'The lower coast resembled Heiberg Island, with mountains and high valleys. The upper coast I estimated as being about one thousand feet high, flat, and covered with a thin sheet ice.'

Cook sent out telegrams claiming he had achieved his original goal: he had reached the North Pole on 21 April 1908 but had been unable to return to Greenland, and so was forced to take refuge on Devon Island. This news shot around the world and he arrived in Copenhagen to a hero's welcome, with the packed audience of his first lecture including members of the Danish royal family.

Then came a peculiar twist. Just five days later, a furious Robert Peary telegrammed from Labrador, claiming that *he* had, in fact, been the first person to reach the North Pole, on 6 April 1909. Peary denounced Cook as a liar, and quoted testimony from the two Inuit, Ahwelaw and Etukishook, that Cook had never even left the mainland. There followed the Cook–Peary Controversy, a public debate as to who had first reached the North Pole that lasted for years, and that to this day has never been fully resolved. Certainly, Cook's case has not been helped by the fact that there is nothing that resembles Bradley Land at the location Cook described.

Then came further mutterings over Cook's reliability: the photographs he provided as proof of his visit to the North Pole were found to be cropped pictures of Alaska that he had taken years before. (His pictures of summiting Mount McKinley a year earlier have also been found to be of an entirely different, and much lower, peak.) He was never able to produce his original navigational records to the pole, and the diary of the expedition he provided to Danish experts for examination had clearly been written much later. The American public, who had previously lauded him as a hero and sided with him over Peary, turned against him, and he travelled the world searching in vain for a safe haven from recognition, often in disguise, which he even adopted to attend a lecture of Peary's in London in 1910. He later became an oil prospector and

formed the Petroleum Producers Association, until he was indicted of mail fraud and sentenced to serve five years in federal prison in Leavensworth.

Deception, one must reluctantly conclude, was his trademark; for it was later revealed by the two Inuit assistants that even the photographs he claimed to have taken of Bradley Land were, in fact, of the coast of Axel Heiberg Island.

Two members of Cook's expedition standing by an igloo holding a US flag – a picture supposedly taken at the North Pole, 1908.

Portrait of Frederick Cook in Arctic furs, 1909.

BUSS ISLAND

58°00'N, 28°00'W

One man for whom the Northwest Passage was a particular obsession was the English seaman Sir Martin Frobisher, who made three voyages in the late sixteenth century in search of a way through the Arctic. During the first of these expeditions, Frobisher landed at a large inlet (now known as 'Frobisher Bay') in the Labrador Sea north of Newfoundland, believing it to be a strait. It was in this area that he discovered a mysterious 'black earth' thought to be rich in ore. A sample the size of a loaf of bread was brought back to England, where the rock was tested by four experts – three dismissed it as mere dirt, while one declared it rich in gold. This was enough to raise support for further voyages, and Frobisher made two more journeys to the Arctic, each time filling his ship with giant loads of the earth – which were eventually proven to be worthless. Though the lack of gold was a disappointment, on the return journey of the third voyage, Frobisher's ship the *Emmanuel* (a sturdy type of vessel known as a 'busse', hence the *Emmanuel*'s sobriquet 'The Busse of Bridgewater') made an altogether different discovery.

A Draught of the Island of Buss, *from John Seller's* English Pilot *(1675).*

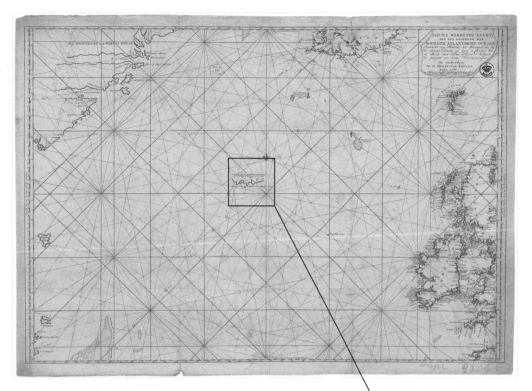

The report of Frobisher's finding was published in George Best's *A True Discourse of the Late Voyages of Discoverie of a Passage to Cathaya* (1570), in which it is written: 'The Busse of Bridgewater, as she came homeward, to the Southeast ward of Freseland [see Phantom Lands of the *Zeno Map* entry on page 240], discovered a great Ilande in the latitude of [erased] Degrees, which was never yet founde before, and sayled three days alongst the coast, the land seeming to be fruitful, full of woods, and a champion countrie.'

This place of 'champion countrie' became known as Buss Island. For fourteen years after this initial discovery, little else is mentioned of Buss, until Richard Hakluyt's *Principal Navigations* appeared in 1598 carrying a description of the island given by a passenger named Thomas Wiars. The details provided by Wiars differ somewhat from those of the *Emmanuel*, describing the island as being largely surrounded by ice and making no mention of its fruits or woods, but adding that it had two harbours at distances of 7 leagues (39km) and 4 leagues (22km) from the southernmost point.

The English navigator James Hall attempted to find Buss while on his way to Greenland in 1605, but failed. On his second attempt, he reported 'a great banke of ice' much further west than Buss was thought to be, but when he tried again

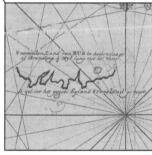

The island of Buss marked on a Dutch naval chart from 1786.

to find the island during his fourth voyage, in 1612, it eluded him once more. Henry Hudson, in 1609, also endeavoured to sight the evasive island but he too failed to find it, although he did report a change in water colour indicating shallow depth. Nevertheless, Hudson remained certain that Buss was real, and the southwestern coast of Buss Island appeared on his chart of the North Atlantic, published in 1612.

The Buss that appeared on the charts in the seventeenth century was an island of considerable size. From north to south, it covered an entire degree of latitude (approximately 69 miles/110km), as it did from east to west, with an unvarying shape. Its omission from several key maps of the period – such as the *New Map* of 1600, the *Map of the World* in Speed's *Prospect* and Hexham's (Mercator's) *Atlas* of 1636 – is indicative of doubt in its existence. Explorers of the area were also sceptical, but, in 1668, there came word that the island had been sighted by a New England sea captain, Zachariah Gillam, during his voyage to Hudson Bay in the ketch *Nonsuch*, who reported land between Iceland and Greenland. Then, on 22 August 1671, Captain Thomas Shepard of the *Golden Lion* (and a former mate on the *Nonsuch*), also on his way to Hudson Bay, claimed to have seen Buss, reporting that 'the Island affords store of Whales, easie to be struck, Sea-horse, Seal and Codd in abundance', and described the island's topography as 'low and level southward … hills and mountains on the northwest end'. This stirred up further interest, and plans were made for an expedition to explore Buss, with Shepard engaged to lead it, but the captain was discharged for 'ill-behaviour' before the voyage got underway and the mission was aborted.

Also convinced of the existence of Buss was the royal hydrographer John Seller, who devoted an entire page in *The English Pilot* (1671) to charting the island, labelling its features with names such as Viner's Point, Rupert's Harbour, Shaftbury's Harbour, Craven Point, Cape Hayes, Robinson's Bay, Albermarle's Point. One finds that no fewer than twelve of these labels derive from the names of the Directors in the Charter of Incorporation granted to the Hudson's Bay Company (HBC), a newly formed company that, in May 1675, was granted rights of trade and commerce and ownership of the island in perpetuity by Charles II. All this preparation was made despite the fact that no one had yet set foot on the island. But it blossomed in the imagination – there was no time to waste, for who knew what kind of natural riches Buss could hold for

Sir Martin Frobisher, c.1535–94.

the man to plant his flag first on its shores? For the sum of £65 the Company received 'the sole trade and commerce of all the Seas Bayes Islettes Rivers Creekes and Sounds whatsoever lying within neare or about the said island … And all mynes Royall as well discovered as not discovered of Gold Silver Gemmes and Precious stones to bee found or discovered within the Island aforesaid.' The HBC dispatched Shepard to Buss with an expeditionary party of two vessels to lay their claim, but unable to find the island they were forced to return empty-handed.

By the eighteenth century, Buss was being treated with deep suspicion. It appeared on several North Atlantic charts, but with maritime traffic increasing in the region the island was conspicuously absent from sightings, and, in 1745, the Dutch cartographer Van Keulen suggested the island had disappeared under the water, with the note: 'The submerged land of Buss is now nothing but surf, a quarter of a mile long, with a rough sea.' The 'Sunken Land of Buss' was then marked on charts as a navigational hazard. This was confirmed in 1791, when Captain Charles Duncan was hired by the HBC to locate the land of Buss, but after a thorough search Duncan struck its death blow, reporting: 'I strove as much as the winds would permit me to keep in the supposed latitude of the supposed Buss island, but it is my firm opinion that no such island is now above water, if ever it was,'

The very rare woodcut map from Luke Foxe's North-West Foxe, or Fox from the North-west Passage … *(1635) showing Buss on the far right.*

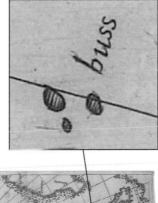

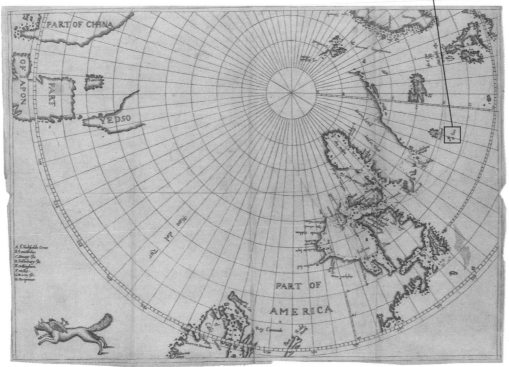

CITY OF THE CAESARS

46°27's, 71°31'55w

Also known as City of Patagonia, Ciudad de los Césares, Elelín, Lin Lin, Trapalanda, Trapananda, the Wandering City

In 1764, there appeared a book, published anonymously, entitled *An Account of the First Settlement, Laws, Form of Government, and Police, of the Cessares, A People of South America*. The remarkable work consists of nine letters written by a J. Vander Neck of Salem, Patagonia, to Mr Vander See of Amsterdam between September 1618 and June 1620, describing a legendary race: the people of the 'Lost City of Caesars'. In the preface, the author attempts to dispel any doubts as to the accuracy of the details within:

How these Letters of Mr. VANDER NECK fell into my hands, it imports the public but little to know. Some of my readers may perhaps view the following account of the Cessares in much the same light with Sir T. MORE'S UTOPIA, rather as what a good man would wish a nation to be, than the true account of the state of one really existing. I shall leave, for an exercise of the Reader's ingenuity, the determination of this point, after only mentioning, that if he pleases to consult Ovalle's Account of Chili in the third volume of Churchill's Collection of Voyages; *Feuillée's* Observations on South America; *and Martinière's* Dictionaire Géographique, *he will find, that there is really a people called the Cessares, in a country near the high mountains, Cordilleras de los Andes, between Chili and Patagonia in South America, in the forty-third or forty-fourth degree of south latitude.*

Seven years after his death, the book was finally attributed to James Burgh, a Scottish educationalist and writer. The letters were completely invented by Burgh and fooled many at the time, in part because his fiction was inspired by a legend that had held a place in the public imagination for hundreds of years. The City of the Caesars was a legendary lost city, believed to be located on an island in the middle of an Andean lake in an area south of Valdivia in Chile, as well as other parts of Patagonia. Its reputation as a place of immense wealth made it a holy grail for treasure hunters, and inspired the same obsessive searching as El Dorado.

Opposite: Bellin's Carte reduite de la partie la plus Meridionale de l'Amerique *(1750), showing the land of the Cezares, north of the 'Pays de Patagons'.*

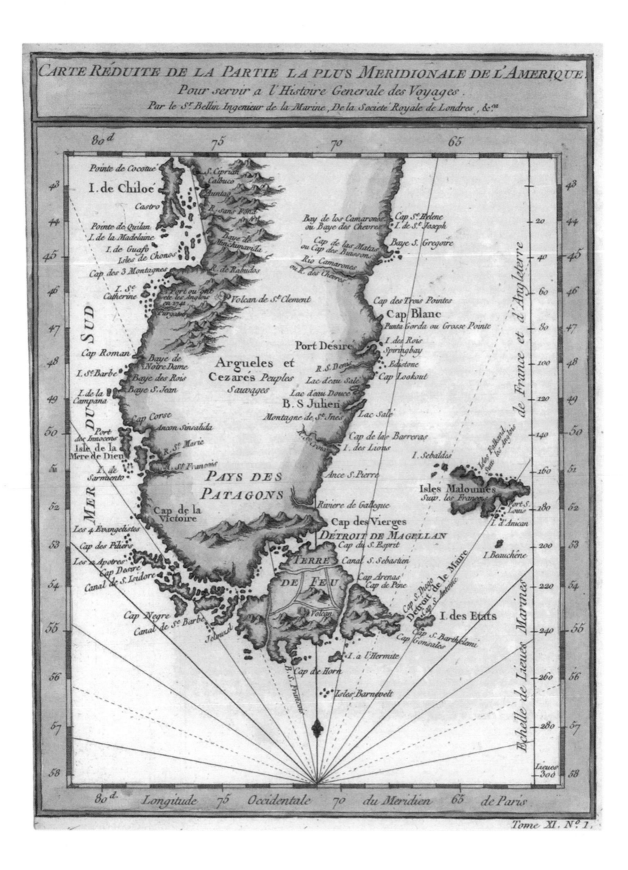

CARTE RÉDUITE DE LA PARTIE LA PLUS MERIDIONALE DE L'AMERIQUE.
Pour servir a l'Histoire Generale des Voyages.
Par le S.ʳ Bellin Ingenieur de la Marine, De la Societé Royale de Londres, &.ᶜᵃ

The myth can be traced back to the voyage of the Italian explorer Sebastian Cabot, who sailed through the Strait of Magellan, the passage between the Atlantic and Pacific oceans off the southern tip of South America, to reach the Mollucas, the spice-rich Indonesian archipelago. In 1528, however, inspired by rumours of a phenomenally wealthy hidden civilization, Cabot sent one of his captains, Francisco César, to lead an expedition party into the deep darkness that was the uncharted South American interior. The party was split into three columns for greater coverage, and entered into the thick jungle – two were never heard from again, most likely because they intruded upon territory of hostile native tribes. César led his group in a northwesterly direction on a three-month journey that covered more than 930 miles (1500km). His official report of the trek is lost; the record we have comes from a Spaniard who passed on the details to an early historian of the

River Plate, claiming to have met César in Peru. Likely a story concocted in this Spaniard's imagination, César was said to have returned to base laden with gold, silver and exotic fabrics, with wild tales of a fabulously wealthy hidden city. If this was indeed the case, then the most likely explanation is that the men stumbled across an outpost of the Incan Empire, although it seems strange that, instead of launching a second expedition to the city, Cabot chose to return to Spain. Regardless, the legend of the City of the Caesars had been brought into being, and caused great public excitement when the story emerged during Cabot's trial in Seville (for diverting from his mission) that his men had seen 'great riches of gold, silver and precious stones', though they were unable to give an exact location.

The intrigue of the Césares was lent especial vibrancy by the frequent disappearances of men of various expeditions in the perilous region throughout the sixteenth century: Simon de Alcazaba's voyage, in 1534, saw a large proportion of the Portuguese expedition abandoned in southern Patagonia; and, in 1540, the 150 men aboard the bishop of Plasencia's flagship were left stranded in the Strait, never to be seen again. One particular report from the Plasencia mission, told to the viceroy of Peru by Cristobal Hernandez and now considered apocryphal, described cities lining a lake 70 leagues (390km) from Córdoba, and of two Spanish survivors who were welcomed into the fold of an Indian tribe, with whom they lived before moving, in 1567, to a fertile land to build a city. It was claimed that these men were the founders of the City of the Césares. The viceroy was won over by the tale and wrote to the king of Spain to request that priests be sent to the area. The search for the city and the lost Spaniards drove several futile expeditions throughout the seventeenth and eighteenth centuries, the last of which was sent by the governor of Chile in 1791. By the mid-nineteenth century, few were left believing in the city's existence, although, like all great legends, its golden lure was sufficient to prevent conclusive dismissal for a good while longer.

SEA MONSTERS OF THE *CARTA MARINA*

Only two known copies exist of Olaus Magnus's hugely imaginative and influential map of Scandinavia, printed in 1539 on nine panels and measuring a total of 49 × 67in (125 × 170cm). The *Carta Marina* is a wonder to behold, its waters teeming with beautiful grotesques – some posing as islands, some shattering ships and some carrying off sailors. To engineer his monstrous aquarium, Olaus took his information from mariners' accounts, medieval bestiaries (such as the *Hortus Sanitatis* of 1485) and popular folklore; and he usefully accompanied each vignette with labels and an elaborative key. Even more helpful was his *Historia de gentibus septentrionalibus* ('A Description of the Northern peoples') printed in Rome in 1555, in which books 21 and 22 provide commentaries on the monsters. Despite the fanciful nature of some of the depictions, Olaus had scientific intentions in presenting an accurate gallery of marine biology – indeed, some of these creatures are recognizable distortions of real animals, while others are purely mythical; but all give an insight into the beliefs and fears that existed in the imagination of the sixteenth-century sailor.

THE ROCKAS

'The benevolence of the fishes called Rockas in Gothic and Raya in Italian: They protect the swimming man and save him from being devoured by sea monsters.'

In his *Historia*, Olaus compares the kindly Rockas, or ray, to the tale told by the German scientist and philosopher Albertus Magnus (*c*.1200), who writes of helpful dolphins that carry swimmers to shore, although he also mentions that, if they suspect the man has ever dined on dolphin flesh, they eat him. Sebastian Münster changes little of the ray in his art, whereas Ortelius gives it the Dutch name 'Skautuhvalur', and jettisons its kindly nature, describing it as: 'completely covered in bristles or bones. It is somewhat like a shark or skate, but infinitely bigger. When it appears, it is like an island, and with its fins it overturns boats and ships.'

part 2. *p.196.*

27. *The great Sea Serpent, according to different Descriptions*

A later depiction of the great sea serpent, from The Natural History of Norway by Erich Pontoppidan, 1755.

THE SEA WORM

'A sea snake, 30 or 40 feet long.'

The coasts of Norway were home to this monster, a blue and grey worm, longer than 40 cubits (about 60ft/18m – 20ft/6m longer than the description in the key to his map), yet as slim as a child's arm. 'He goes forward in the Sea like a Line, that he can hardly be perceived how he goes; he hurts no man, unless he be crushed in a mans hand: for by the touch of his most tender Skin, the fingers of one that toucheth him will swell.' The animal – which sounds very much like an exaggerated eel – had a natural enemy in the crab, the strong pincers of which it could not escape. 'I oft saw this Worm', wrote Olaus, 'but touched it not, being fore-warned by the Marriners.'

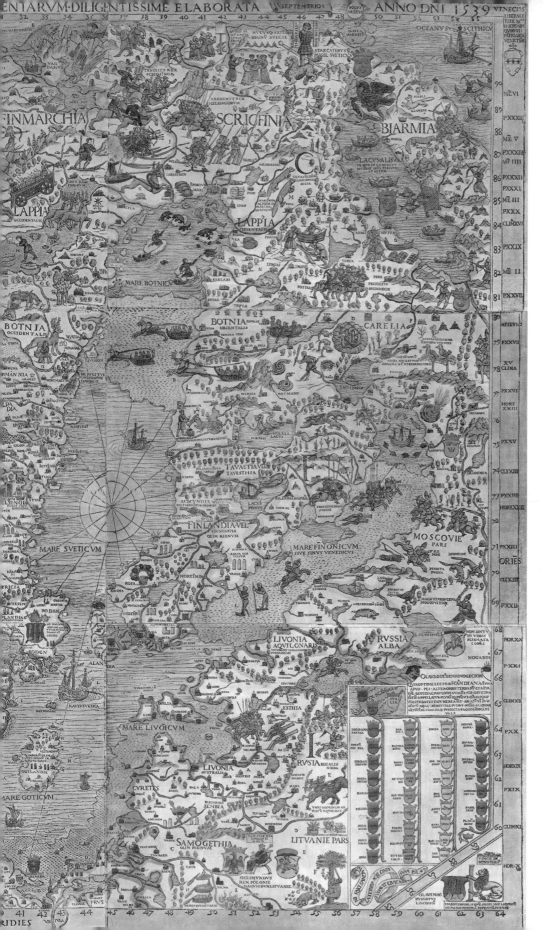

Olaus Magnus's Carta marina et description septemtrionalium terrarum ac mirabilium *(Nautical Chart and Description of the Northern Lands and Wonders) (1527–39).*

THE DUCK TREE

'Ducks being hatched from the fruit of trees.'

This mythical plant, said to sprout baby birds, seemed to explain the breeding of ducks, a mysterious affair for it was conducted when the birds flew south. Of the Sollendae ducks, often seen near Glegorn, Scotland, Olaus writes: 'Moreover, [a] Scotch Historian, who diligently sets down the secret of things, saith that in the Orcades [Orkney Isles], Ducks breed of a certain Fruit falling into the Sea; and these shortly after get wings, and fly to the tame or wild Ducks.' This is a variation on the 'barnacle goose tree' myth, of which the archdeacon and historian Gerald of Wales wrote in the twelfth century:

Enclosed in shells of a free form they hang by their beaks as if from the moss clinging to the wood and so at length in process of time obtaining a sure covering of feathers, they either dive off into the waters or fly away into free air … I have myself seen many times with my own eyes more than a thousand minute corpuscles of this kind of bird hanging to one log on the shore of the sea, enclosed in shells and already formed … Wherefore in certain parts of Ireland bishops and religious men in times of fast are used to eat these birds as not flesh nor being born of the flesh.

Barnacle Geese growing on a tree, depicted on a medieval manuscript.

THE POLYPUS

'A Polypus, or creature with many feet, which has a pipe on his back.'

This giant lobster viciously preyed on mariners and swimmers 'with his Legs as it were by hollow places, dispersed here and there, and by his Toothed Nippers, he fastneth on every living creature that comes near to him, that wants blood. Whatever he eats, he heaps up in the holes where he resides: Then he casts out the Skins, having eaten the flesh, and hunts after fishes that swim to them.' The Polypus could change his colour to blend in with his environment, something he did to escape his most feared enemy, the conger eel.

BALENA AND ORCA

'A whale, a very great fish, and the Orca, which is smaller, his deadly enemy.'

A Whale is a very great fish about one hundred or three hundred foot long, and the body is a vast magnitude; yet the Orca, which is smaller in quantity, but more nimble to assault, and cruel to come on, is his deadly Enemy. An Orca is like a Hull turned inside outward; a Beast with fierce Teeth, with which, as with the Stern of a Ship, he rends the Whales Guts, and tears his Calves body, or he quickly runs and drives him up and down with his prickly back, that he makes him run to the Fords, and Shores.

The whales can be seen here by the island of Tile, thought to be Thule (see Thule entry on page 230).

THE SEA PIG

'A sea monster similar to a pig.'

Now I shall revive the memory of that monstrous Hog that was found afterwards, Anno 1537, in the same German Ocean, and it was a Monster in every part of it. For it had a Hogs head, and a quarter of a Circle, like the Moon, in the hinder part of its head, four feet like a Dragons, two eyes on both sides of his Loyns, and a third in his belly inkling toward his Navel; behind he had a Forked-Tail, like to other Fish commonly.

Like several of Olaus's depictions, the Sea Pig derives from the observations of Pliny, who described a 'pig-fish' that grunted when it was caught. Most likely, what is being described here, fantastically, is the walrus.

THE SEA UNICORN

The basis here is, of course, the narwhal, the large tusks of which were often found washed up on beaches. 'The Unicorn is a Sea-Beast, having in his Fore-head a very great Horn, wherewith he can penetrate and destroy the ships in his way, and drown multitudes of men. But divine goodnesse hath provided for the safety of Marriners herein; for though he be a very fierce Creature, yet is he very slow, that such as fear his coming may fly from him.'

THE PRISTER

'The Whirlpool, or Prister, a kind of whale whose floods of waters sink the strongest ships.'

The species of whale is unidentified, but the description of spouting appears similar to the Balena:

The Whirlpool, or Prister, is of the kind Whales, two hundred Cubits long, and is very cruel. For to the danger of Sea-men, he will sometimes raise himself beyond the Sail-yards, and casts such floods of Waters above his head, which he had sucked in, that with a Cloud of them, he will often sink the strongest ships, or expose the Marriners to extream danger. This Beast hath also a long and large round mouth, like a Lamprey, whereby he sucks in his meat or water, and by his weight cast upon the Fore or Hinder-Deck, he sinks and drowns a ship.

Olaus advises scaring it away with a 'Trumpet of War', as it can't bear the sharp noise. Failing this, he says, cannons should do the trick.

THE ZIPHIUS

'The terrible sea monster Ziphius devouring a seal.'

Though its name comes from *xiphias,* the Greek word for sword, this creature is completely separate to what we know as a swordfish. The blade of this owl-faced monster appears to be the sharp dorsal fin on its back:

Because this Beast is conversant in the Northern Waters, it is deservedly to be joined with other monstrous Creatures. The Sword-fish is like no other but in something it is like a Whale. He hath as ugly a head as an Owl: His mouth is wondrous deep, as a vast pit, whereby he terrifies and drives away those that look into it. His Eyes are horrible, his Back Wedge-fashion, or elevated like a sword; his snout is pointed. These often enter upon the Northern Coasts, as Thieves, and hurtful Guests, that are always doing mischief to ships they meet, by boaring holes in them, and sinking them …

THE SEA-COW

Along with descriptions of the Sea-mouse, the Sea-hare and the Sea-horse are provided details of the Sea-cow, drawn identically to the land animal: 'The Sea-Cow is a huge

Monster, strong, angry, and injurious; she brings forth a young one like to her self; yet not above two, but one often, which she loves very much, and leads it about carefully with her, whither soever she swims to Sea, or goes on Lands … Lastly, this Creature is known to have lived 130 years, by cutting off her tail.'

THE SEA RHINOCEROS

Olaus only makes reference to this spotted creature in the key to his map, which states: 'A monster looking like a rhinoceros devours a lobster which is 12 feet long.' With such scant information, it has been suggested that this could well be the cartographer succumbing to a bout of *Horror vacui,* and filling a space with something of his own creation.

THE ISLAND WHALE

'Seamen who anchor on the back of the monsters in belief that they are islands often expose themselves to mortal danger.'

Olaus's whale has a substance on its skin similar to seaside gravel, and so, when it raises its back above the waters, sailors are tricked into thinking the mound is an island. When they reach it, they climb its 'shore', drive piles into the surface to which they attach their ships and then kindle fires to cook their meat. The whale, feeling the fire, immediately dives down to the bottom, and all on his back, unless they can save themselves by ropes thrown from the ship, are drowned.

Whales of such size that they are mistaken for islands and mountains are quite common in early literature, from Sinbad's first voyage in the *Arabian Nights*, to the fourth-century *Physiologus* (thought to be the source of Olaus's inspiration here), in which sailors cast their anchor into a giant 'Aspidoceleon'. In this account, though, the monster is a tool of religious symbolism – the beast is Satan, and readers are warned that: 'if you fix and bind yourself to the hope of the devil, he will plunge you along with himself into the hell-fire.' (See also St Brendan's Island entry on page 202.)

Olaus's sea creatures had significant influence over later works of other cartographers, including this famous map of Iceland by Ortelius from 1590.

THE SEA SERPENT

'A worm 200ft (60m) long wrapping itself around a big ship and destroying it.'

There is a Serpent which is of a vast magnitude, namely 200 foot long, and more – over 20 feet thick; and is wont to live in Rocks and Caves toward the Sea-Coast about Berge … He hath commonly hair hanging from his neck a Cubit long, and sharp Scales, and is black, and he hath flaming shining eyes. This Snake disquiets the Shippers, and he puts up his head on high like a pillar, and catcheth away men, and he devours them; and this hapneth not but it signifies some wonderful change of the Kingdom near at hand; namely that the Princes shall die, or be banished; or Tumultuous Wars shall presently follow.

This is the first written account of the sea orm, or Norway serpent. Perhaps, it is influenced by the story of Jörmungandr ('Great Beast'), a 'Midgard Serpent' from Norse mythology, that grew so large in the depths of the ocean that it eventually wrapped itself around the world.

CARIBDIS

'Several horrendous whirlpools in the sea.'

'Here is the horror Caribdis' reads the label on the map accompanying this monster, an ancient myth famously featured in *The Odyssey*, *Jason and the Argonauts* and Aristotle's *Meteorologica*. Here Olaus draws a ship caught in the terrible whirlpool, and writes:

Wherefore those that would sail thither from the Coasts of Germany hire the most experienced Marriners and Pilots, who have learned by long Experience, how by steering obliquely, and directing their course … they may not fall into the Gulph … Also the Sea there, within the hollow Cave, is blown in when the Flood comes, and when it ebbs, it is blown out, with as great force as any Torrents or swift Floods are carryed. This Sea, it is said, is sailed in with great danger, because such who sail in an ill time are suddenly sucked into the Whirl-pools that run around.

ISLAND OF CALIFORNIA

30°00'N, 115°10'W

European explorers dreamed of the Californian utopia before any had stepped foot on its shores or even confirmed its geography. The fantasy can be traced to a popular Spanish novel published in 1510 by García Ordóñez de Montalvo, called *Las Sergas de Esplandián*, in which the author writes:

Know that to the right of the Indies there is an island called California very close to the side of the Earthly Paradise; and its inhabitants were black women, without a single man, for they lived in the manner of the Amazons. They were beautiful and their bodies robust, with fiery courage and great strength. Their island is the most formidable in the world, with its steep cliffs and stony shores. Their weapons are all made of gold, as are the harnesses they use to tame their wild beasts, because there is no other metal on the island other than gold.

La Californie ou Nouvelle Caroline … *by Nicholas de Fer (1720) shows one of the largest and finest depictions of the island of California ever produced.*

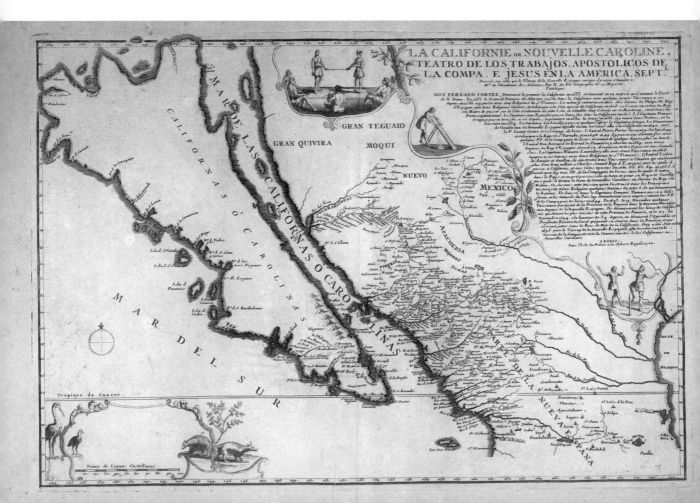

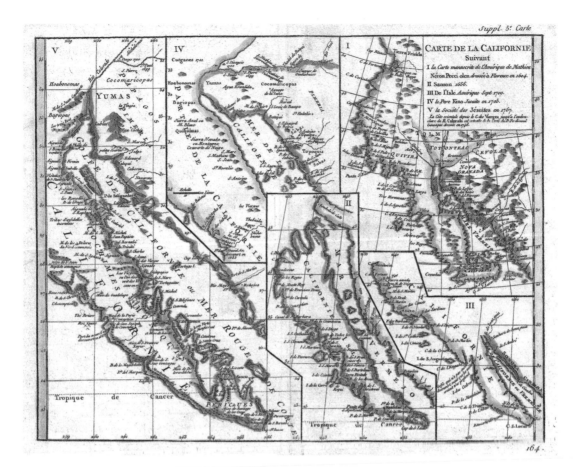

*Robert de Vaugondy engraved
this in 1770 for the* Denis
Diderot Encyclopedie,
*illustrating the various confused
states of the island of California.*

The myth drove Hernán Cortés, the Spanish conquistador who
brought about the fall of the Aztec Empire, to send expeditions
to find the Amazonian island: in 1533, a party led by his cousin
Diego de Becerra and Fortun Ximenez landed on the southern
tip of Baja California peninsula, believing it to be surrounded
by sea on all sides. Their report reached Cortés, who sent
further explorations: Francisco de Ulloa tracked the coast
northward until, reaching the Colorado river, he discovered
the island to be a peninsula; this was then confirmed by the
navigator Hernando de Alarcón.

Soon, California began to appear on maps, making its debut
on a 1541 drawing by Domingo del Castillo, shown correctly as
part of the mainland. Then, in 1562, it appeared on a printed
map for the first time as part of Diego Gutiérrez's portrayal of
the New World. Mercator and Ortelius reproduced it on their
works, and for sixty years the Californian peninsula enjoyed
accurate representation.

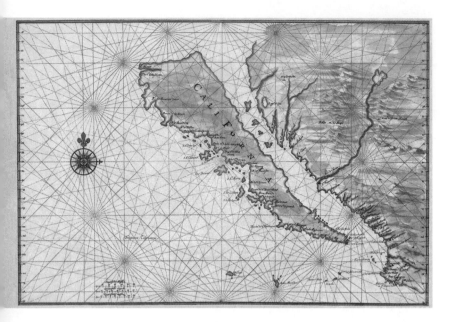

Johannes Vingboon's Map of California as an Island, *c.1650.*

But then something strange happened – California was redrawn as an island. The first to render its newly divorced state was Michiel Colijn of Amsterdam, on the title page of *Descriptio Indiae Occidentalis* in 1622. The island misconception was then reproduced as a matter of course for decades: by Abraham Goos in 1624; by John Speed in 1627; by Henry Briggs in 1625; and Richard Seale in 1650. In fact, 249 maps showing California Island (not including world maps) were identified by the historians Glen McLughlin and Nancy H. Mayo in 1995. For the entire seventeenth century, and for most of the eighteenth, cartographers wrenched California free from the American continent and set it adrift in the Pacific Ocean.

The mythical reinvention is thought to have originated from the 1602 voyage of Sebastian Vizcaino up the Californian coast, an account of which was written twenty years later by the Carmelite friar Antonio de la Ascensión, who had been on board Vizcaino's ship. In this journal, Ascensión describes California as being separated from the mainland by the 'mediterranean Sea of California'. His descriptions were mapped and issued to Spain, but the vessel carrying the records was hijacked by the Dutch, and the misinformation was accepted and adopted by their publishers.

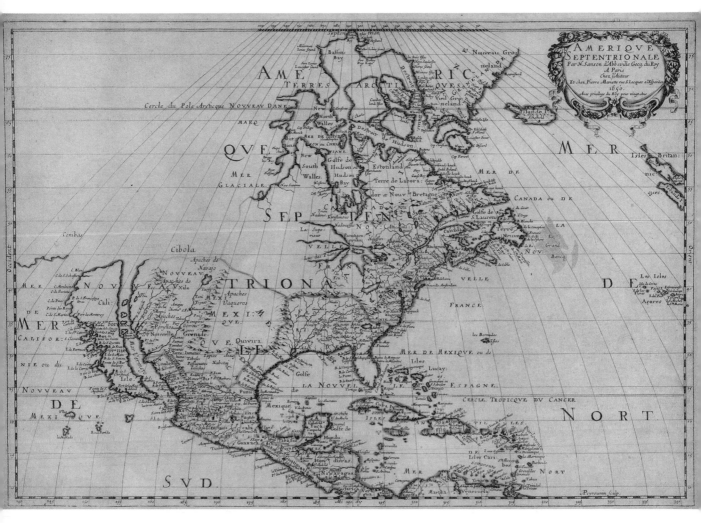

Major cartographers such as Willem Blaeu and Herman Moll fell for the blunder, and lent it credence with their own reproductions. It wasn't until 1706 that doubt began to be cast. The Jesuit friar Eusebio Kino, who was initially a believer in the island notion, made a series of journeys from Sonora to the Colorado river delta. His realization that it was connected to the mainland is reflected in the map accompanying his personal accounts. Further confirmations were made, and eventually Ferdinand VI of Spain was moved to issue an official decree in 1747, declaring: 'California is not an island.' The reports of Juan Bautista de Anza, from his 1774 travels between Sonora and the west coast of California, effectively reattached the island to the mainland, although, strangely, it makes one curious, much later appearance on a Japanese map by Shuzo Sato in 1865.

Island of California shown on a map by Nicolas Sanson.

CASSITERIDES

50°19'N, 8°13'W

Also known as the Tin Islands; the Cassiterida

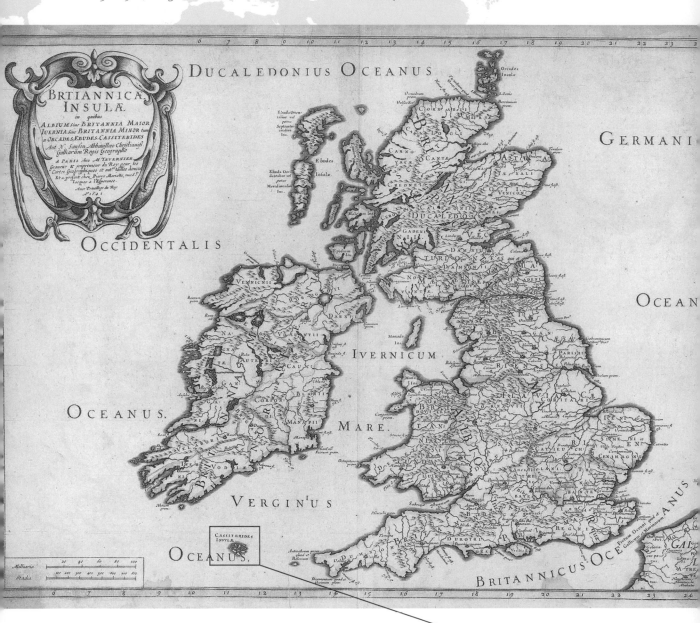

To the Ancient Greeks the Cassiterides, or 'Tin Islands', were the mysterious source of their tin and lead, a collection of islands located somewhere deep in murky western Europe, their specific position a jealously guarded secret of the Phonenicians, who dominated the metal trade at the time.

Herodotus makes mention of them but admits no knowledge of their position; while Diodorus describes them as islands of tin 'lying in the ocean over against Iberia'. Strabo (64/63 BC –AD *c*.24) provides a few further details, describing them in *Geographica* (3.5.11) as being ten in number, and grouped north of the haven of the Artabri (a tribe living in Galicia in northwest Spain). One of the islands is a desert, he states, but the others are inhabited by men in black cloaks reaching down to their feet and fastened at the chest, who walk with staves and resemble the Furies from tragic representations. These islanders live off their cattle, and for the most part lead a wandering life. 'Of the metals they have tin and lead; which with skins they barter with the merchants for earthenware, salt, and brazen vessels.'

It is also Strabo who describes Rome's encroachment on the Phoenician trade, mentioning an incident in which the Romans attempted subtly to follow a certain shipmaster in order to discover the source of his goods. On noticing the pursuing ship, the captain deliberately ran his vessel upon a shoal, leading the Romans to do the same with disastrous results. The shipmaster, however, escaped by floating away on a fragment of his vessel, and received from the state the value of the cargo he had lost.

There have been many theories as to the true identity of the Cassiterides, including: the Cornwall region of Great Britain, the Scilly Isles off the southwest coast of Britain and the British Isles as a whole – as well as Spain and its surrounding islands. Tin-rich Britain is certainly more likely than Spain – a conclusion clearly shared by the French cartographer Nicolas Sanson, who, on the map opposite, depicts the British Isles in the time of the Roman Empire and makes a rare inclusion of the Cassiterides, which appear to be a slightly relocated Scilly Isles. This interpretation of the classical sources as drawn by Sanson was certainly popular, but it is Roger Dion's proposal in *Le problème des Cassitérides* (1952) that has been suggested as most likely. He describes the former existence of a number of islands off France's west coast in a wide gulf beyond the Bay of Biscay, before it silted up, where now one finds the marshes of the Brière, between Paimboeuf and St Nazaire. This would also seem to work with Strabo's description of the Cassiterides being ten islands in the sea, north of the land of the Artabri in northwest Spain.

Opposite: Nicholas Sanson's 1694 map of Great Britain, showing the Cassiterides isles off the southwest coast.

CROCKER LAND

83°00'N, 100°00'W

In 1906 the American explorer Commander Robert
Edwin Peary was making a bid to be the first to reach
the North Pole, driving his dogsled relentlessly through
biting winds across the rough terrain of the frozen Arctic
Ocean. Standing on the northwest summit of Cape
Thomas Hubbard, he paused to brush the ice from his
eyes and glimpsed an enormous mass of land glittering
in the distance. 'My heart leaped the intervening miles
of ice as I looked longingly at this land', he wrote later
in *Nearest the Pole* (1907), 'and in fancy, I trod its shores
and climbed its summits.' He named it Crocker Land,
in honour of the San Francisco banker George Crocker,
who had contributed $50,000 to Peary's expedition, and
built a stone cairn on the spot, leaving inside a written

*Map of the Crocker Land
expedition from the* New York
Tribune, *11 May 1913.*

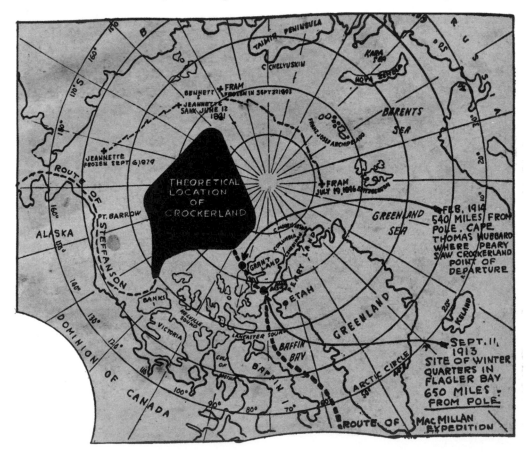

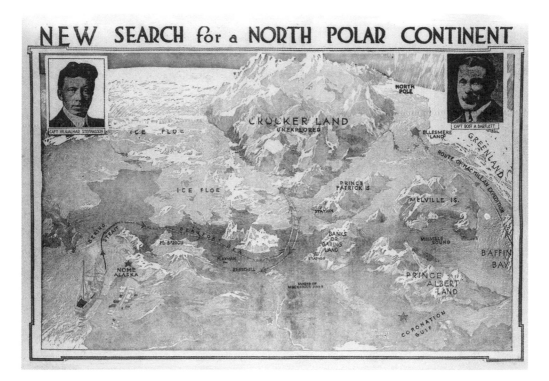

NEW SEARCH for a NORTH POLAR CONTINENT

account of his visit. By Peary's estimation, the land was approximately 130 miles (210km) from the cape at roughly the coordinates given above. However, there is no such island, or indeed any other land formation in that region. Possibly Peary was tricked by a looming type of mirage, but, damningly, in his original diary there is no mention of sighting Crocker Land – in fact, the entry of 24 June 1906 reads: 'No land sighted'. From this, it seems he inserted the discovery later in his journal out of design to flatter Crocker and secure funding for his next attempt to reach the North Pole.

That expedition was launched in 1908 and, as mentioned in the story of Bradley Land (see relevant entry on page 42), involved Peary becoming entangled in a furious argument with Frederick Cook as to who was the rightful discoverer of the pole. As it turned out, the existence of Crocker Land became a key issue in the Peary–Cook debate – for, in his account, Cook claimed to have passed through the coordinates given by Peary for Crocker Land and found that no such place existed. Those backing Peary's claim, therefore, determined to find the land mass and prove Cook to be a liar once and for all.

The San Francisco Call *painted this scene for its readers on 27 July 1913.*

Robert Peary in furs, 1907.

    The American explorer and Peary's former lieutenant
Donald MacMillan organized the expedition. MacMillan
had started his career as a high school teacher, leading a
seamanship and navigation summer camp. During one season
he saved nine people from wrecked boats, a story that came
to the attention of Peary, who invited him to join his 1905
effort to reach the North Pole during which their friendship
was forged. For the mission to defend Peary's honour,
MacMillan secured funding from the American Museum
of Natural History (which raised the modern equivalent of
1 million dollars from industrialist backers for the mission), the
American Geographical Society and the University of Illinois.
A large number of donors were members of the Peary Arctic
Club of New York. He also enlisted a small team of academic
experts to accompany him. Serving as guide and translator
was Minik Wallace, one of six Inuit whom Peary had brought

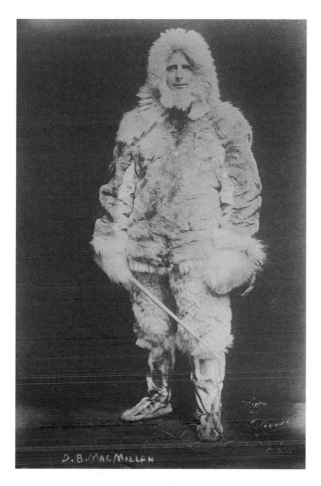

Donald MacMillan, c.1910.

home with him from his Arctic voyage in 1897.* Describing the Crocker Land question as 'the world's last geographical problem', MacMillan announced the mission to the world at a press conference in 1913:

In June 1906, Commander Peary, from the summit of Cape Thomas Hubbard, at about latitude 83 degrees N, longitude 100 degrees W, reported seeing land glimmering in the northwest, approximately 130 miles [240km] away across the Polar Sea. He did not go there, but he gave it a name in honor of the late George Crocker of the Peary Arctic Club. That is Crocker Land. Its boundaries and extent can only be guessed at, but I am certain that strange animals will be found there, and I hope to discover a new race of men.

---

* *As well as returning with several Inuit, Peary also stole two giant iron meteorites ('the Woman' and 'the Dog') from the Inuit, who used the rocks to make their tools. The rocks are today housed in the American Museum of Natural History.*

On 2 July 1913, the men
departed from the Brooklyn
naval yard aboard a steamer
bound for Greenland. Ill
fortune struck the expedition
two weeks into the voyage: the
steamer's captain got drunk
and steered the ship into an
iceberg, wrecking the vessel.
The explorers eventually
transferred to another ship
named the *Erik*, and continued
on their way, landing at north
Greenland in August.

After making the necessary
preparations, on 11 March
1914 the team of MacMillan, a
fresh-faced 25-year-old Navy
ensign named Fitzhugh Green,
biologist Walter Ekblaw, seven
Inuit (who were paid with
rifles and biscuits) and their
125 sled-dogs embarked on the
1200-mile (1930km) trek across
the polar ice to find Crocker
Land, fighting through fierce
storms and temperatures that
plummeted below -22°F (-30°C). The party eventually reached
the Beitstadt Glacier, and spent three days scaling its 4700ft
(1433m). The temperature dropped further: Ekblaw suffered
extreme frostbite and was taken back to the base camp by
several of the Inuit guides. As MacMillan doggedly pushed on,
other members of the team abandoned the mission, until finally
by 11 April only he, Green and the Inuits Piugaattoq and
Ittukusuk were left. Together they crossed the frozen Arctic
Ocean on sleds, until on 21 April MacMillan cried out that he
could see Crocker Land. 'There could be no doubt about it,' he
wrote later in his memoir. 'Great Heavens! What a land! Hills,
valleys, snowcapped peaks extending through at least one
hundred and twenty degrees of the horizon.'

Their experienced guide Piugaattoq calmly explained to
the American that what he was seeing was, in fact, a common
mirage called 'poo-jok', meaning mist. MacMillan ignored the

*Minik Wallace in New York in
1897. He was brought to America
with his father and other Inuit
by Robert Peary, to be studied by
staff of the American Museum
of Natural History, which had
custody. After his father died,
museum staff tricked Minik
into thinking they had given his
father a proper burial – instead
they put his skeleton on display
to the public.*

native – he had found proof of his friend's claim! – and gave the order to continue over the treacherous, breaking ice. For five days more the men chased the band-shaped mirage (now thought to be a Fata Morgana), until MacMillan was forced to admit that they were pursuing an illusion. He wrote:

The day was exceptionally clear, not a cloud or trace of mist; if land could be seen, now was our time. Yes, there it was! It could even be seen without a glass, extending from southwest true to northeast. Our powerful glasses, however … brought out more clearly the dark background in contrast with the white, the whole resembling hills, valleys and snow-capped peaks to such a degree that, had we not been out on the frozen sea for 150 miles, we would have staked our lives upon its reality. Our judgment then as now, is that this was a mirage or loom of the sea ice.

The men turned and headed for land. Concerned that the changing weather would isolate them from their camp, MacMillan ordered Green to take Piugaattoq and search for an alternative route to the west. As the two set off, the weather descended and forced the men to take shelter in a snow cave. For the young and inexperienced Green, the situation was terrifying. As the storm claimed one of their dog teams the cabin pressure intensified. An argument broke out, and in a fit of pique Green took a rifle from the sled and murdered Piugaattoq, shooting him in the back for, he said, failing to follow orders.

When Green rejoined MacMillan and the others on 4 May, he confessed what had happened, but asked that the Inuit be told that Piugaattoq had been killed by the storm. Green was never prosecuted for the murder, even though it was suspected that there was an ulterior motive to the crime, for it was rumoured that he had developed a sexual relationship with Piugaattoq's wife (who had also borne two children to Peary).

Due to adverse conditions, the expedition was then stranded in northern Greenland for a further three years until its members were finally able to return to America with abundant anthropological research, furs, photographs, specimens and blood on Green's hands – though nothing to support the existence of Peary's Crocker Land. Ekblaw described the episode as 'one of the darkest and most deplorable tragedies in the annals of Arctic exploration'.

CROKER'S MOUNTAINS

74°22'N, 94°02'W

In the early nineteenth century, the confusion caused
by an illusory mountain range spotted off the eastern
coast of Greenland led to acrimonious debate, the public
ridicule of a respected British naval officer, and a serious
delay in uncovering the Northwest Passage.

In 1818, three years had passed since the Napoleonic Wars
and the British fleet lay idle in dock. This offered a chance
to pursue non-martial preoccupations. Reports were coming
in from whalers that the ice packs to the east of Greenland
were breaking up at an unprecedented rate, and so, under
Sir John Barrow, Second Secretary to the Admiralty, efforts
were renewed to hunt for a long-sought trade route through
the Arctic to Asia. The command of the first expedition was
handed to 41-year-old John Ross, a capable Scot who had
joined the British Navy aged nine as an apprentice, and spent

*A chart of the track of Ross's
expedition, taken from his
A Voyage of Discovery
(1818). Croker's Mountains
are drawn on the far west
side, apparently walling up
Lancaster Sound.*

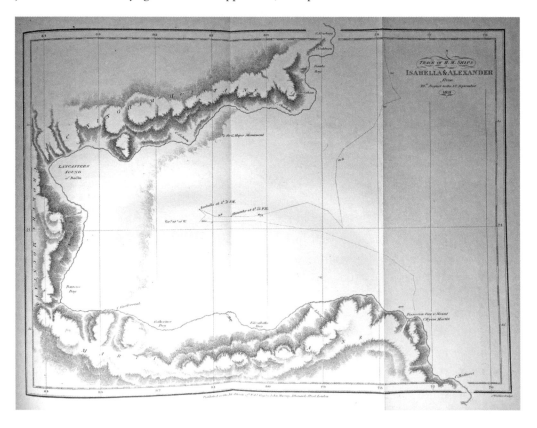

Drawn by Capt. Ross, R.N.   Engraved by R. Havell & Son.

John Ross's sketch of his
expedition's passage through
the ice in June 1818.

the following thirty-two years developing an illustrious naval
career, including a captaincy in the Swedish Navy.

In April 1818, Ross and his crew sailed the flagship *Isabella*
down the Thames to the cheers of crowds, and set off to find
the passage. Following close behind in the consort *Alexander*
was his second-in-command, Lieutenant William Edward
Parry, thirteen years his junior but an expert in Arctic climates
from his experiences of the previous few years guarding the
whale fisheries of Spitsbergen (now known as the Norwegian
archipelago Svalbard).

After a brief stop in the Shetland Islands, the men sailed on
for Greenland. The first leg of the voyage was relatively
uneventful: previous discoveries were reconfirmed, and various
scientific measurements of tides, ice and magnetism were taken.
Ironically, considering what was to come, Ross and his crew even
disproved the existence of the sunken land of Buss, passing
directly over its location (see Buss Island entry on page 46).

The *Isabella* and the *Alexander* then reached Baffin Bay off
Greenland's southwest coast. They circled in an anticlockwise
direction, methodically corroborating observations made of
the area by William Baffin two hundred years earlier, and
made contact with tribes of Inuit along the northwest coast
(referred to by Ross as 'Arctic Highlanders'). On 30 August,
they reached, and entered, the inlet that Baffin had named
Lancaster Sound. As they sailed between Devon Island and
Baffin Island their excitement grew, for it seemed as though

Ross's sketch of Croker's Mountains.

they had discovered a gateway to the Northwest Passage (which, indeed, they had). This was swiftly dashed when, after several miles of easterly sailing, Ross observed there to be a mountain range ahead that apparently sealed off Lancaster Sound as a bay. This range, which he named 'Croker's Mountains' after John Wilson Croker, the First Secretary of the admiralty, would obstruct any further progress. His officers were baffled: they insisted there was no such mountain range ahead, that what he was seeing was a mirage – they must sail on! To their utter frustration, Ross stubbornly ignored both the protests of his men and the weight of responsibility to push forward and find the passage, and made the extraordinary decision to turn the ship around and abandon the mission: 'To describe our mortification and disappointment', wrote the outraged purser aboard the *Alexander*, 'would be impossible at thus having our increasing hopes annihilated in a moment, without the shadow of a reason appearing.'

There is, indeed, no such mountain range at the position of Croker's Mountains. In his published account of the expedition, *A Voyage of Discovery* ... (1818) which was written hastily upon his return to defend himself from the fierce criticism, John Ross describes the sighting:

I distinctly saw the land, round the bottom of the bay, forming a connected chain of mountains with those which extended along the north and south sides. This land appeared to be at the distance of eight leagues; and Mr. Lewis, the master, and James Haig, leading man, being sent for, they took its bearings, which were inserted in the log; the water on the surface was at temperature of 34. At this moment I also saw a continuity of ice, at the distance of seven miles,

extending from one side of the bay to the other, between the nearest
cape to the north, which I named after Sir George Warrender, and
that to the south, which was named after Viscount Castlereagh.
The mountains, which occupied the centre, in a north and south
direction, were named Croker's Mountains, after the Secretary to
the Admiralty. The southwest corner, which formed a spacious bay,
completely occupied by ice, was named Barrow's Bay ...

Ross even included an exculpatory sketch of the mountains he saw.

On their return to England, Ross's officers continued
furiously to contest his mountain sighting. Most vehement was
William Parry, who was certain Lancaster Sound was a strait
(and who was later given command of HMS *Hecla* for his
own – more successful – expedition in 1819). In particular, Ross
came under especially harsh criticism from Sir John Barrow,
who later commented on Ross's first expedition with an acidic
entry in his own book *Voyages of Discovery and Research within
the Arctic Regions, from the Year 1818 to the Present Time* (1846):

Among the little irregularities of Commander Ross, it can not
escape notice that he addresses all his letters and orders issued during
the voyage, and unnecessarily printed in his book, as from John
Ross, captain of the Isabella. His promotion to that rank on his
return was easily acquired, being obtained by a few months' voyage
of pleasure round the shores of Davis's Strait and Baffin's Bay,
which had been performed centuries ago, and somewhat better, in
little ships of thirty to fifty tons. It is a voyage which any two of the
Yacht Club would easily accomplish in five months.

Ross was pilloried for his apparent stupidity – or worse –
cowardice. In a bid to salvage his reputation, he launched
a second Arctic expedition in 1829. Out of necessity it was
privately funded, underwritten by the London gin magnate
Felix Booth, who contributed £17,000 to the £3,000 that
Ross himself put up. (Accordingly, when Ross and his men
discovered a new peninsula in the northern Canadian Arctic at
70°26'N, 94°24'W, they named it 'Boothia Felix'.) In contrast
to the relatively easy tour of his first voyage, this second
expedition was fraught with problems. Ross and his men had
to dump their experimental boiler engine overboard and press
on by sail alone, and they ended up trapped for four frozen
winters in the region as their captain searched relentlessly, and
in vain, for the vindication of finding a passage through the ice-
packed landscape.

DAVIS LAND

27°12's, 91°22'w

Today, the term 'buccaneer' is generally interchangeable for pirate, but in the seventeenth century it was a specific label for the men who attacked and plundered Spanish shipping and settlements in the Caribbean and along the Pacific South American coastline.* The buccaneers operated with a remarkable impunity, for they were regarded by English authorities as an unofficial extension of anti-Spanish operations, and were granted 'letters of marque', which authorized them to capture enemy ships and bring them home for sale.

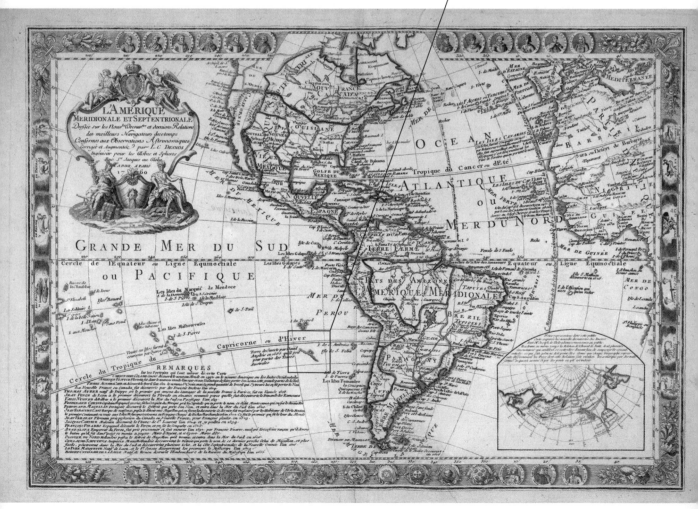

The shift in tactics from sacking ships to launching mainland assaults was marked in 1654 by the attack on New Segovia in Honduras, the first in a brutal campaign of pillaging by notorious privateers including Henry Morgan, a Welshman with a ruthless reputation. In 1655, Morgan's ship was commissioned by the governor of Jamaica, Sir Thomas Modyford, to help reinforce the security of his island. The privateers, however, took these orders to be a carte blanche to indulge in havoc. They attacked a series of littoral towns, seized the islands of Providencia and Santa Catalina, northwest of Colombia, and launched an all-out assault on Panama. So drunk on their successes were they that Morgan struggled to keep his men from invading Peru.

The opportunity for unbridled violence drew others to the area in search of riches. After a brief period of peace, in 1680 Panama was assailed once more, by a company of buccaneers led by Peter Harris. They, however, were driven back and instead commandeered a fleet of ships and under the new leadership of the pirate captain Bartholomew Sharpe embarked on an eighteen-month voyage of coastal attacks around the Pacific. When Sharpe eventually decided to return home to Britain, he inadvertently became the first Englishman to round Cape Horn in an easterly direction, after a storm blew him off course.

Among the ensemble of cut-throats and sea thieves that made up Sharpe's crew were two men of particular note: Lionel Wafer and William Dampier. Both were unusual in that they were well-educated men. They had originally met in Cartagena, before joining Sharpe's crew in 1680, though their experiences were decidedly different. After a quarrel, Wafer was abandoned with four others in the Panamanian 'Isthmus of Darien', where he proceeded to live among the Cuna Indians for a year, studying their culture and learning their language. Dampier, meanwhile, sailed on. After Wafer was eventually reunited with the buccaneers – who initially failed to recognize their friend dressed in local costume – he and Dampier eventually ended up aboard the *Bachelor's Delight*, captained by Edward Davis. It is of this voyage that

The buccaneer and explorer William Dampier clasping his best-selling book, A New Voyage Round the World.

Opposite: J. B. Nolin illustrated Dampier's 'long tract of pretty high land' on his map of L'Amérique *(1760), with a French annotation that translates as 'land discovered by David Anglois in 1685, that he took for part of Terres Australes'.*

---

* *The word 'buccaneer' originates from the cooking technique adopted from the Caribbean locals by the Europeans. The French verb* boucanier *('to cook or cure') comes from the Caribbean Arawak word* buccan *('a wooden frame for smoking meat – usually manatee').*

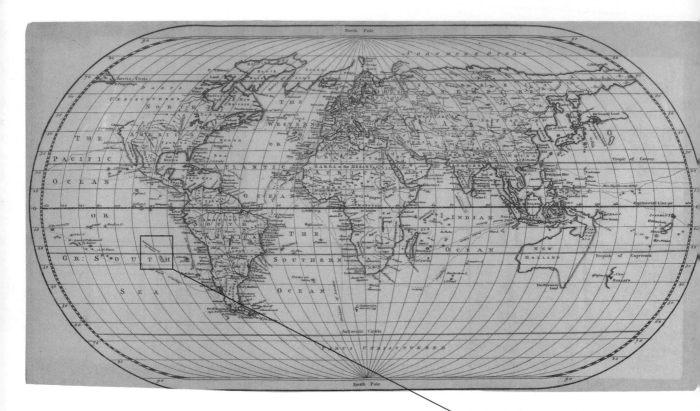

both Dampier and Wafer later published accounts of their travels on board. Though heavily romanticized, and sprinkled with fictitious details (as these journals so often are), Dampier's *A New Voyage Round the World* (1697) and Wafer's *A New Voyage and Description of the Isthmus of America* (1699) both make reference to a particular mystery: the discovery of Davis Land, 1500 miles (2780km) off the coast of Chile.

Sailing southward, several hundred leagues from the South American coast and at 12° latitude (the same parallel as Lima), and after enduring a sea-earthquake, the crew of the *Bachelor's Delight* sighted what was to be claimed as Davis Land. Wafer writes:

Having recov'rd our Fright, we kept on to the Southward. We steer'd South and by East, half Easterly, until we came to the Latitude of 27 Deg. 20 Min S. when about two Hours before Day, we fell in with a small, low, sandy Island, and heard a great roaring Noise, like that of the Sea beating upon the Shore, right a Head of the Ship. Whereupon the Sailors, fearing to fall foul upon the Shore before Day, desired the Captain to put the Ship about, and to stand off till Day appeared; to which the Captain gave his consent. So we plied off till Day, and then stood in again with the Land; which proved to be a small flat Island, without the guard

Map from Henry Ellis's Considerations on the Great Advantages which would arise from the Discovery of the North West Passage … *(1750).*

of any Rocks. We stood in within a quarter of a Mile of the Shore,
and could see it plainly; for 'twas a clear Morning, not foggy nor
hazy. To the Westward, about 12 Leagues by judgment, we saw
a range of high Land, which we took to be Islands, for there were
several Partitions in the Prospect. This Land seem'd to reach about
14 or 16 Leagues in a range, and there came thence great Flocks of
Fowls. I, and many more of our Men would have gone ashore at it;
but the Captain would not permit us. The small Island bears from
Copayapo almost due E. 500 Leagues; and from the Gallapago's,
under the Line, 600 Leagues.

While not aboard the ship with Wafer during that period,
Dampier also mentions the discovery of the Pacific Island,
as an anecdote later told to him by Davis:

Captain Davis told me lately that, after his departure from us at
the haven of Realejo (as is mentioned in the 8th chapter) he went,
after several traverses, to the Galapagos, and that, standing thence
southward for wind to bring him about Tierra del Fuego in the
latitude of 27 south, about 500 leagues from Copayapo on the coast
of Chile, he saw a small sandy island just by him; and that they saw
to the westward of it a long tract of pretty high land tending away
toward the north-west out of sight. This might probably be the
coast of Terra Australis Incognita.

Herman Moll represented Davis Land on a world map frontispiece
in Dampier's *New Voyage* and, within a year of Dampier's
account being published, Davis Land had also appeared on
numerous French maps. J. B. Nolin illustrated Dampier's 'long
tract of pretty high land' on his map of *L'Amérique* in 1740,
with the annotation 'Terre découverte par David Anglois.'
(The French insisted on referring to Davis as David.)

Several ventures in search for Davis Land were also inspired
by the accounts of Wafer and Dampier (see Pepys Island entry
on page 186 for more on these two). The Dutch West India
Company dispatched three ships to the area in 1721 under the
command of Jacob Roggeween. Though unable to find the
island, Roggeveen and his crew stumbled across a hitherto
unknown land mass that did exist – Easter Island. It is thought
that this was the land mass spotted by Davis and his crew,
and because of some erroneous calculations they wrongly
recorded their actual position. Nevertheless, towards the end
of the eighteenth century, British and French ships were still
searching in vain for Davis Land.

ISLE OF DEMONS

55°11'N, 49°19'W

Also known as Île des Démons, I. dos Demonios

Strange noises were said to emanate from the Isle of Demons, a dark land of legend once believed to be Quirpon Island in Newfoundland, Canada. 'A great clamor of men's voices, confused and inarticulate', reports André Thevet, the French explorer priest who published an account of his experiences on the 'Isola des Demonias'. The place was rumoured to be home to evil spirits and terrible creatures that viciously attacked those foolish enough to enter its waters or tread its soil. Thevet claimed to have survived by defending himself with the Gospel of St John.

An early appearance of the Isle of Demons can be found on a 1508 map of Johannes Ruysch, who shows two 'Insulae Demonium' near the centre of the Ginnungagap passage between Labrador and Greenland. This passage was greatly feared by Norsemen, probably because of the existence of maelstroms. Though Satanic Islands in the North Atlantic are a common feature of early medieval folklore (in part because of the strong storms brought by the Arctic current), it is thought that the Demon Islands could be a relocated depiction of the older island 'Satanazes', moved north of another fictional land, Antillia, by cartographers when charting of the original area was improved and no island was found. (Some legends are just too intriguing to expunge altogether; see Satanazes entry on page 210 and Antillia entry on page 18.) Sebastian Cabot mapped them as a single island, 'Y. de Demones', in 1544, shifting it nearer to the eastern front of Labrador, Newfoundland. Gerardus Mercator put them by the upper tip of Newfoundland, while Ortelius maintains Cabot's position for the single island on his map of 1570.

The island is the setting for one particular folktale concerning a pregnant aristocrat named Marguerite de la Roque. When her affair with a sailor was discovered, Marguerite, her lover and her servant-girl Damienne were abandoned on the Isle of Demons by her uncle, Jean-François de la Roque de Roberval, and left to the mercy of the beasts of the island. Marguerite gave birth, but the child, the sailor

and the handmaiden all died, so Marguerite was left alone,
fending off wild animals with firearms and surviving by her
wits until by chance she was rescued by Basque fishermen.
She returned to France and was, for a time, made famous by
the story, before settling down as a schoolmistress in Nontron.
André Thevet, the sixteenth-century French geographer, wrote
of the story in *La Cosmographie universelle* (1575), claiming to
have heard it from the heroine herself whom he encountered
at Nontron in Périgord; but, in fact, he was just repeating the
account in *Heptaméron* (1558) by Marguerite de Navarre (who
said she heard it from 'Captain Roberval') and in François de
Belleforest's *Histoires tragiques* (1570). The story was also the
subject of a poem by the nineteenth-century Canadian poet
George Martin, entitled *Marguerite, or the Isle of Demons*.

The Isle of Demons is found on late sixteenth-century maps
by Mercator, Ruysch and Ortelius, but by the mid-seventeenth
century had been removed from official cartographic records.

*Map by Thomas Jefferys, taken
from Theodore Swaine Drage's*
The Great Probability of a
North West Passage *(1768).*

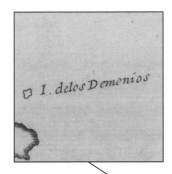

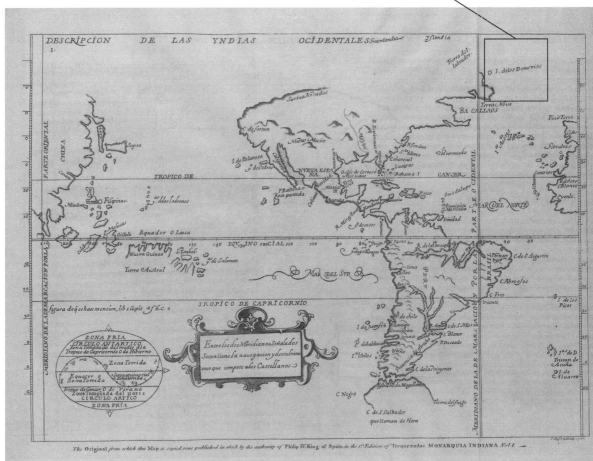

HAVANA Portus St. DOMINGO. CARTAGENA. MEXICO. CUSCO.

Groenlandi.

Virginiani.

Rex et Regina Floridæ.

Nova Albionis Rex.

Mexicani.

AMERICA

SEP

TENTRIONALIS

FRAN
CIA

Septentrionalissimas Americæ par=
tes, Groenlandiam puta, Islandiã
et adjacentes, quod Americæ tu=
bulæ commodè comprehendi
non potuerint, peculiari hac ta=
bella Spectatoribus exhibendas
duximus.

Tropicus

Cancri

MAR

DEL

ZUR

CIRCULUS ÆQUINOCTIALIS

OCEANUS

PERUVIANUS

Tropicus Capricorni

MARE PACIFI=

CUM

AMERICÆ
nova Tabula.
Auct. Guilielmo Blaeuw.

Cum privilegio
dicem annorum.

TERRA AUSTRALIS INCOGNITA.

Willem Blaeu's
Americæ nova
Tabula, *showing the*
'I. dos Demonios', to
the west of phantoms
'As Maydas' and
'Brazil'; and at the
top of South America
can be found the
mythical Lake Parime,
supposedly the site of
El Dorado; and, at the
very south, a pair of
Patagonian Giants.

DOUGHERTY ISLAND

59°20's, 120°20'w

In May 1931, Captain Mackenzie, master of the exploratory vessel *Discovery*, triumphantly announced before a group of reporters that he and his crew had slain an island:

Today the Discovery passed over the charted position of Dougherty Island, that mysterious outpost of the Great White South, which has been described as one of the most desolate and isolated spots in the ocean, and which has for so long evaded the search of expedition ships, whalers, and other vessels. Good observations were obtained throughout the morning, and at noon our course was altered to enable us to pass directly over the Island's assigned position, where, at 2 p.m., soundings revealed a depth of 2470 fathoms. No land was seen, and there are no indications of its existence in this locality. The weather for once has been remarkably clear, and had there been an island within 12 miles of this spot we could not have failed to see it.

Dougherty Island made its first official appearance on record in 1841, when Captain Dougherty, skipper of the English whaler *James Stewart*, documented an encounter with an island in the southernmost part of the Pacific Ocean, 5–6 miles (8–9.5km) in length with a high bluff to the northeast end and lower land beyond covered in snow. This was then given credence by connecting it with a report from 1800 of 'Swain's Island' spotted in the same general vicinity. So far, the tale of Dougherty varies little from that of many other phantom islands – the result of a mirage, perhaps, or a captain confused by an especially large iceberg. But what makes Dougherty Island intriguing is the fact that it was then repeatedly encountered by a number of navigators, who provided almost identical coordinates and descriptions. Captain Keates of the *Louise*, for example, met the island in 1860, even mentioning an iceberg caught on its shore, and gave its coordinates as 59°21'S, 119°07'W. On his return, the calibration of his chronometer was examined, and found to be accurate within ¼ mile (0.4km).

In 1891, Dougherty Island was sighted by Captain Stannard, the master of the English barque *Thurso*, belonging to Lyttelton, New Zealand, who reported falling in with the

Opposite: Map of the Antarctic from a German atlas of 1906 published by Justus Perthes. (Also shows Maria Theresa Rock; and Sannikov Land – see relevant entries on pages 156 and 208.)

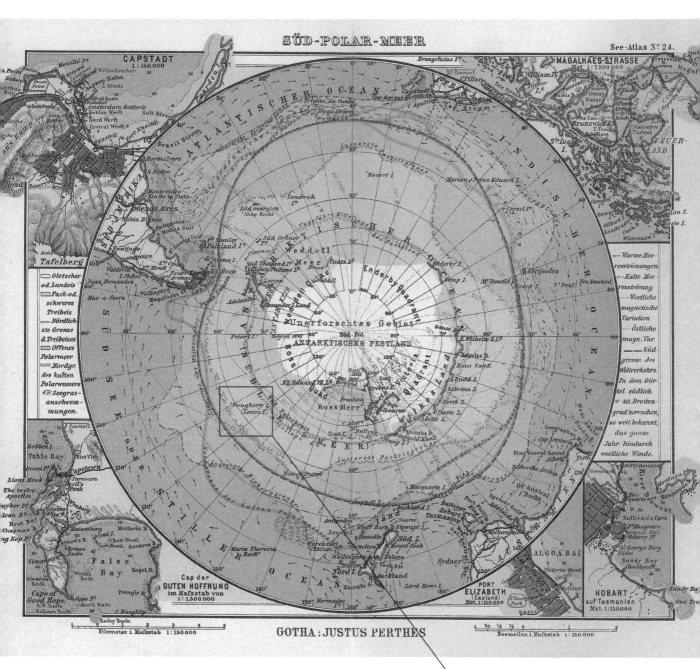

See-Atlas N.º 24.

CAPSTADT
1 : 150.000

MAGALHÃES-STRASSE
Mst. 1 : 7.500.000

ATLANTISCHER OCEAN

INDISCHER OCEAN

ANTARKTISCHER

Weddell Meer

Enderby Quadrant

Unerforschtes Gebiet

ANTARKTISCHES FESTLAND

Süd Pol

Ross Quad.

Victoria Quad.

Wilkes Land

Ross Meer

Dougherty I.
(Keates I.)

SÜD-SEE

STILLER OCEAN

Cap der
GUTEN HOFFNUNG
im Maſsstab von
1 : 1.500.000

Table Bay

False Bay

Cape of
Good Hope

ALGOA BAI

PORT
ELIZABETH
(Capland)
Mst. 1 : 150.000

HOBART
auf Tasmanien
Mst. 1 : 150.000

Kilometer i. Maſsstab 1 : 150.000

GOTHA : JUSTUS PERTHES

Seemeilen i. Maſsstab 1 : 150.000

island on his voyage to England from New Zealand, marking it at 59°21'S, 119°07'W. As he approached the region, he checked his charts and found the area to be drawn clear; but, seeing as the weather was such that he had to remain in the neighbourhood during the night, he kept a sharp lookout, and sure enough he sighted an island, and was close to it for part of three days. He reported the island with a description remarkably similar to that of Captain Dougherty – it being

Dougherty I.
(Keates I.)

about 6 miles (10km) long, with a bluff on the northeast end about 300ft (90m) high, quite free from snow. The captain mentioned passing several icebergs before sighting the island, to prove his ability to differentiate land and ice.

By this point, Dougherty Island was established on charts of the vicinity, and many expeditions set out to do what none so far had managed: to land on its shore. Between 1894 and 1910, Captain Greenstreet of the British steamer *Ruapehu* made five separate attempts at finding it, while the Antarctic explorer Robert Falcon Scott searched in vain for it in 1904. In 1909, John King Davis, captain of Ernest Shackleton's vessel *Nimrod*, conducted a search for several lost islands, including Dougherty. His journals were included in an appendix in the second volume of Shackleton's account of his second expedition, *The Heart of the Antarctic* (1911):

The 1909 journey of the SY Nimrod, *captained by John King Davis, searching for lost islands in the South Pacific Ocean. Taken from Davis's paper published by the Royal Geographical Society, 1910.*

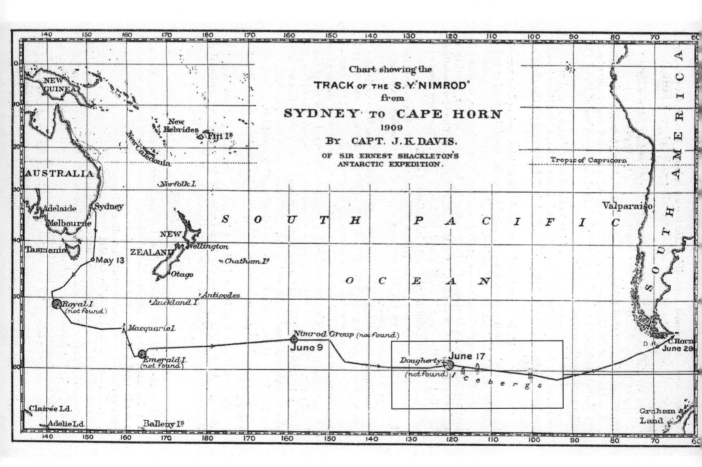

At noon on June 17 we were up by dead reckoning to the position of Dougherty Island, as given by Captain Dougherty, but as the weather was overcast could not be sure of our position. Captain Keates places the island in the same latitude thirty-four miles further east. I therefore continued eastward on the parallel over this position (by dead reckoning). As it was now dark and the weather moderate, I stood back again to the westward, hoping to get sights at daylight and did so. Good star observations were confirmed at noon, when the island, according to Captain Dougherty's position, should have borne west distance four miles. No land was in sight from the masthead in clear weather. I stood east again, and at 4 p.m., when darkness was just setting in, the island according to Captain Keates' position should have borne east four miles; nothing in sight. At 4.30 we passed over this position and continued eastward along the parallel of 59°21' South, but saw no indication of land. It is just here that we met with ice during our passage, and I am inclined to think Dougherty Island has melted.

Definitive disproof, one would have thought. But, for as many refutations, there are testimonies to its existence. One of these was mentioned by the oceanographer Henry Stommel, who wrote in 1984 about an indignant letter in the file on Dougherty Island at the Admiralty Hydrographic Department. The author is a Mr St Clair Whyte of Auckland, New Zealand, who protests the attempt to expunge Dougherty Island from the map, claiming to have recently spent several hours clubbing seals offshore close to its charted position and betting 'all the tea in China' that it exists there.

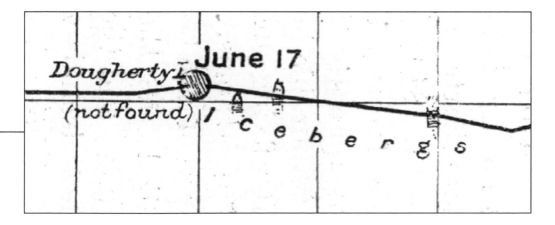

EARTHLY PARADISE

In the Welsh language there is a particularly beautiful word: 'Hiraeth'. It has no direct English translation, but the general sense of the term is an overwhelming feeling of grief and longing for one's people and land of the past, a kind of amplified spiritual homesickness for a place one has never been to. This word came to mind when collecting together maps drawn over the centuries by Europeans attempting to give a specific earthly location for an unearthly concept – the Garden of Eden. For it is one of the most ancient and alluring of all myths: man's first and perfect home, lost and lying somewhere out there over the seas, waiting to be rediscovered; a terrestrial microcosm of a heavenly realm in some desert oasis or on a remote island in the distant East.

The notion can be found in varied forms in religions around the world. The Greeks had their 'Golden Age', a mythical primordial era of blissful harmony watched over by the titan Cronus. The Earth provided man with everything he needed and people lived to a grand old age, becoming guardian spirits when they died. The Chinese, Babylonians, Egyptians and Sumerians, too, have their equivalents. For the Christians, the Garden of Eden (originally an ancient Hebrew tradition) was where God 'drove out the man; and he placed at the east of the Garden of Eden Cherubims, and a flaming

The Psalter Mappa Mundi, *created c.1265, depicting Adam and Eve just below Jesus at the head of the map.*

sword which turned every way, to keep the way
of the tree of life' (Genesis 3:24). Eden is described
as the source of four giant rivers: Pishon, Gihon
(Nile), Hiddekel (Tigris) and Euphrates. This has
similarities to the legend of Mount Meru in Jainism,
Hinduism and Buddhism, which is believed to be
man's ancient home, the seat of gods, and a place
where four rivers burst forth into the cosmic ocean.

The essential imagery of a four-rivered garden
somewhere in the farthest East is how one finds
the story's earliest graphic form on maps. On
the medieval *mappae mundi*, which are usually
oriented with east as north, Jesus is found at the
top, presiding over a paradoxical combination of
legend and geography. The example presented here
is the *Psalter World Map*, so called because it was
discovered in a book of psalms. Though relatively
small in size, it is one of the great medieval maps
of the world (though probably not the original,
more likely a copy of a work long since lost that
once hung in the Westminster Palace bedchamber
of Henry III around the mid-1230s). Just like the
famous *Hereford Mappa Mundi*, its detail provides
insight into contemporary understanding of
ancient history, scripture and geography. One of
these vignettes is a painting of the Eden tale – the
portraits of Adam and Eve sit just below Jesus, atop
their four rivers.

In 1406, a Byzantine manuscript of Ptolemy's
Geography, generally thought to have been lost
to Western Europe until this time, was sent from
Constantinople to Venice and, after the publication
of Jacopo Angelo's translation of the second-century
work into Latin, it began to have a profound effect
on European cartography. The development of
sea charts had already introduced out of necessity
the practical purpose – and, therefore, scientific
approach – of mapping, moving away from the
historical display of *mappae mundi*. But it was in
the fifteenth century that the east-oriented layout

The Garden of Earthly Delights. *Part of
a triptych by Hieronymus Bosch, c.1500.*

of the *mappae mundi* popularly gave way to a preference for the Ptolemaic model, the basis of modern mapping, with the introduction of a mathematical system using global coordinates, and the establishment of north at the head of the map.

This presented a problem for Renaissance mapmakers – belief features, such as the Earthly Paradise, had no place on maps that prized accuracy over religious decoration – and so the garden began to be dropped from maps, its abandonment a symbol of the progression in contemporary thinking as

The Ptolemaic system – 1482 world map from Nicholaus Germanus's Cosmographia.

God introduces Adam and Eve to the Garden of Eden in a painting by Lucas Cranach, c.1530.

the sixteenth century dawned. Consequently, attempts were made to rationalize the myth: scholars such as Vadianus and Goropius Becanus argued that the garden could be interpreted not as a specific tangible territory but as Adam and Eve's pure and blissful existence before the arrival of sin. Martin Luther found it pointless to argue about the exact location of the garden, as it was probably destroyed with everything else in the Great Flood, a victim of Man's sin.

The French theologian John Calvin agreed with the idea that the garden had been drowned and lost, but presented the comforting theory that God maintained affection for Man and had left remnants of the paradise on Earth. Calvin accompanied his *Commentary on Genesis* (1553) with a map of Mesopotamia with its rivers, and claimed the garden to have once been in the region. He interpreted the 'four rivers' to mean four 'heads' of rivers, that is, two channels carrying the water to the garden, and two bearing it away, and showed how this could fit the Mesopotamian system. Calvin's idea was adopted and developed by various sixteenth-century publishers

Map of Paradise in Mesopotamia, *from Thomas Guarinus's* Biblia Sacra *(1578).*

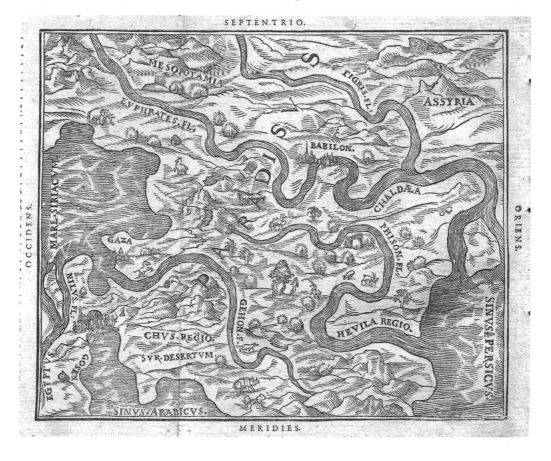

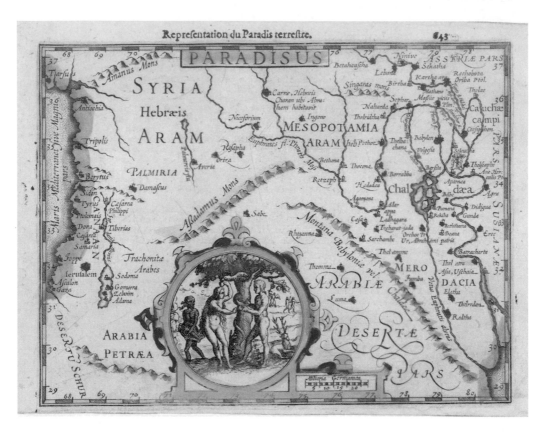

of the Bible, some of whom clarified the idea for their readers and also went a step further, drawing the site of the Fall on the map. One example from Thomas Guarinus's *Biblia Sacra*, from 1578, is shown on page 95.

In 1607, Mercator and Hondius developed this design with greater geographic detail for the map *Paradisus*, which is dominated by a vignette of Adam and Eve below the apple tree.

The four-river design of the garden was dropped from maps soon after. One of the last to feature the emblem was Sir Walter Raleigh, who, a few years after Hondius, included the Earthly Paradise on his map of *Arabia the Happie*, together with a scattering of other biblical imagery. The idea that the Garden had been obliterated by the Flood continued to be accepted, but from this point there was a shift in theory as to its original location: attention turned from Mesopotamia to Armenia, which, at this time, included the region between the Upper Euphrates and Lake Urmia, the Black Sea and the Syrian desert. Perhaps, it was thought, the biblical river Pishon was, in fact, the River Phasis, and the Gihon the River

Map from 1607 by Mercator and Hondius, showing Paradise near Babylon.

Opposite top: A map by Romain de Hooge, c.1700, with the Garden of Eden depicted in the centre.

Opposite bottom: A late map of the Garden of Eden *by Pierre Mortier (1700) was based on the theories of Pierre Daniel Huet, Bishop of Avranches. The caption reads: 'Map of the location of the terrestrial paradise, and of the country inhabited by the patriarchs, laid out for the good understanding of sacred history, by M. Pierre Daniel Huet.'*

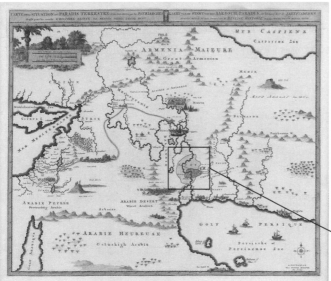

Araxes. After this, the Holy Land was offered as another
alternative, but this was more to do with dogmatic convenience
than any geographic indication. By the eighteenth century, the
cartography of the Earthly Paradise was, for most, a decoration
of antiquity, and it was left to its perennial verdancy in
imagination and religious myth.

EL DORADO

3°09′N, 58°09′W

Also known as Manoa

Gold has always been the reason. Myths and rumours of auric lands and cities have lured men blindly over oceans, into uncrossable deserts and the deep 'green hell' of impenetrable jungles. European obsessions with the riches of the East are as old and powerful as imagination – the glittering kingdom of Prester John, studied later, epitomized these foreign fantasies – yet golden ghosts date back to antiquity. Ptolemy, for example, describes an 'Aurea Regio' found in 'India beyond the Ganges'; and, in the mid-first century, *The Periplus of the Erythraean Sea* (a list of ports and coastal landmarks written in Greek) referred to 'Chryse', the Land of Gold, describing it as 'an island in the ocean, the furthest extremity towards the east of the inhabited world, lying under the rising sun itself, called Chryse … Beyond this country … there lies a very great inland city called Thina.' (Dionysius Periegetes, *fl*. CE 120, also writes of 'The island of Chryse, situated at the very rising of the Sun'.) In the fourth century the Roman writer Avienus refers to the Insula Aurea (Golden Isle) as being found where 'the Scythian seas give rise to the Dawn'.

When Columbus first arrived in Latin America in 1492, it was with dismay that he discovered the country was not the Spice Islands of Asia that he so desperately sought. But, by the mid-sixteenth century the New World of the Americas had delivered greater riches than had been imagined in the form of Aztec and Inca treasure and the bountiful native mines. At the time Columbus had embarked for the Indies, the sum total of Europe's gold would have formed a cube of a 'mere' 8 cubic yards (6 cubic metres). From 1503 (the year Columbus concluded his fourth voyage) to 1560, Spain's portion of this was greatly exceeded by a vast haul totalling 101 tons of gold that was shipped home from the New World. The historian Fernand Braudel calculated that by the middle of the sixteenth century Spanish reserves of gold and silver rocketed to a modern

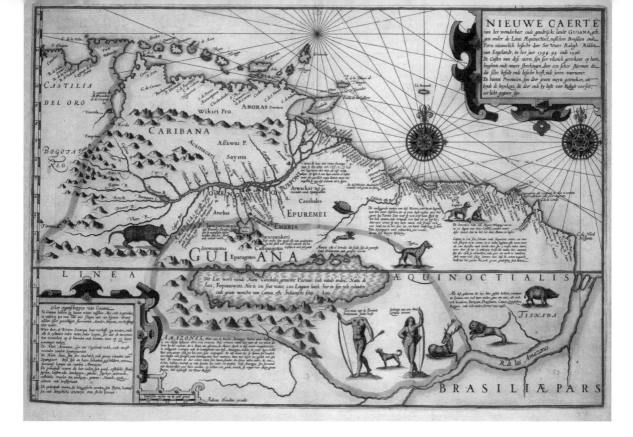

value of more than 2 trillion dollars. 'Gold is the most excellent of metals', writes Christopher Columbus in a letter to King Ferdinand dated 7 July 1503. 'With gold we not only do whatever we please in this world, but we can even employ it to snatch souls from Purgatory, and to people Paradise.'

Of course, they would want more. Surely, pushing further into the vast jungle heartland would result in even greater discoveries. There were whispers of lost cities of gold in the Andes (fame passed on, no doubt, by natives aware that the lure of gold elsewhere was the way to rid one of the Spanish). The rumoured Inca city of Paititi, ruled over by the emperor of Musus, was said to lie hidden somewhere in the rainforests on the modern borders of Bolivia, Brazil and Peru. (Regular efforts to find Paititi continue to this day.)

The most beguiling legend, though, was the story of The Gilded One, or 'El Dorado'. The title of this city centred on the daily ritual of its king, which was described by Gonzalo Fernández de Oviedo y Valdés in *La Historia General de las Indias* (1535):

When I asked why this prince or chief of king was called Dorado, the Spaniards who had been in Quito and had now come to San Domingo (of whom there were more than ten here) answered, that,

Hondius's map Nieuwe caerte van het Wonderbaer ende Goudrjcke Landt Guiana *(1598) of Sir Walter Raleigh's description of the region now known as French Guyana. As well as the mythical Lake Parime, marked as the site of El Dorado/Manoa, there are the Blemmye (headless men) in the south as well as fanciful depictions of South American fauna.*

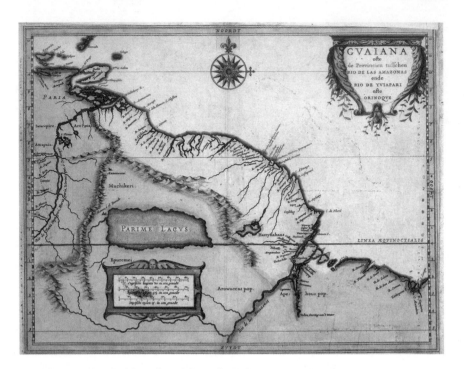

according to what had been heard from the Indians concerning that great lord or prince, he went about constantly covered with fine powdered gold, because he considered that kind of covering more beautiful and noble than any ornaments of beaten or pressed gold … [He] put his on fresh every morning and washed it off in the evening … The Indians further represent that this cacique, or king, is very rich and a great prince, and anoints himself every morning with a gum or fragrant liquid, on which the powdered gold is sprinkled and fixed, so that he resembles from sole to crown a brilliant piece of artfully shaped gold.

Map of Guaiana by Hessel Gerritz (1625) showing Lake Parime, on the shores of which the city of El Dorado was said to be found.

The story was apparently based on a practice of the Muisca people, who lived high in the mountains of New Granada. Where was this tribe, so rich as to think nothing of discarding gold daily? Sebastián de Belalcázar, conqueror of Nicaragua, set off to find out in late 1535. When he finally reached the Muisca, he discovered the area had been settled already, by Gonzalo Jiménez de Quesada, who had found disappointingly little in the way of gold; meanwhile reports came in that the German adventurer Nicholas Federmann was also nearby, scouring the mountains, drunk on the same stories. The lake in which the king washed off his gold was key to the myth. The Spanish believed it to be Lake Guatavita, and, beginning in 1540 with a party led by Hernán Pérez, repeated efforts were made to drain it, but no gilt lining was found.

The last organized Spanish search for El Dorado was made by a battle-hardened group led by the bloodthirsty Lope de Aguirre, a Spanish conquistador who took his command by murder, which came to an end when the maniac led his rag-tag bunch on all-out invasion of Venezuela. As the Spanish became convinced of the city's non-existence, and the Germans also grew dispirited, along came the British to pick up the golden goose chase. From prisoners captured at Port of Spain, Sir Walter Raleigh learnt the story of The Gilded One. In *The Discovery of Guiana* (1595), which is a celebrated example of fantastical geography and myth propagation, he conflates the El Dorado myth with that of another lost city, Manoa:

I have been assured by such of the Spaniards as have seen Manoa, the imperial city of Guiana, which the Spaniards call El Dorado, that for the greatness, for the riches, and for the excellent seat, it far exceedeth any of the world, at least of so much of the world as is known to the Spanish nation. It is founded upon a lake of salt water of 200 leagues long, like unto Mare Caspium.

This he wrote as being located on the legendary Lake Parime, far up the Orinoco River in Venezuela. It sits alongside other enjoyable inaccuracies in the text, including Raleigh's description of the Ewaipanomas, a tribe of headless men with facial features in their chests. (This particular story dates back to antiquity – see Creatures of the *Nuremberg Chronicle Map* entry on page 179.)

From this point, the dazzling shine of El Dorado dwindled to a fraction of its sixteenth-century wattage, though not before Raleigh persuaded King James I to back one last mission to find his Manoa in 1617. The voyage was a disaster: his ships were smashed by storms, and his men – many failing to share his vision – deserted their commander. Raleigh, weakened by a tropical illness, attacked the Spanish outpost of Santo Tomé de Guayana on the Orinoco River and in the ensuing fight his son was killed. When he returned to England, heart-broken and empty-handed, the Spanish demanded Raleigh be punished for breaking the peace treaty that he had expressly promised to obey. The treason charge for his involvement in the Main Plot (a 1603 conspiracy to remove James I from the throne), for which he had earlier been pardoned, was revived, and for his reckless pursuit of the golden city Sir Walter Raleigh was beheaded in the Old Palace Yard at the Palace of Westminster on 29 October 1618.

Sir Walter Raleigh, engraved by Robert Vaughan in 1650.

FLAT EARTH

This map was created by a gentleman named Professor
Orlando Ferguson of Hot Springs, South Dakota, USA
in 1893 to illustrate his firm belief that the Earth was not
spherical, as those in the scientific community would
have one believe, but flat and square. Ferguson and his
fellow refuters based their alternative geodesy on literal
interpretation of passages from the Bible: take, for example,
the quote from Revelations 7:1 (which also adorns the
home page of The International Square Earth Society

Professor Orlando Ferguson's
Map of the Square and
Stationary Earth *(1893).*

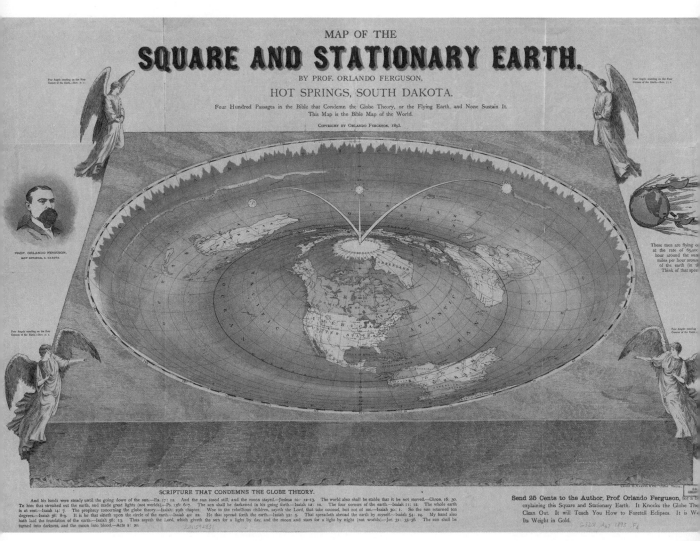

website at the time of writing): 'After this I saw four angels standing at the four corners of the earth, holding back the four winds of the earth to prevent any wind from blowing on the land or on the sea or on any tree.'

'Four corners' of a terrestrial plain, taken literally, was held up as proof of the world's flatness. And then there is the description in Isaiah 11:12 of 'the four quarters of the Earth' – quarters was interpreted as suggesting four symmetrical parts, which limits options of shape to square, rectangular and similar variations. Finally, the Revelations quote is drawn upon again with the same stubborn internal logic: the angels stand at the four corners of a quartered Earth, each holding back one of the four winds – these are known to be the north, south, east and west winds, the four opposite points of the compass. Therefore, the angels, maintaining the quartered symmetry, must be facing each other equidistantly. The only shape that fits this is a diamond or a square and, since God is infallible, the perfect square is *therefore* the shape that matches.

As for his claim of the Earth being stationary and not in orbit around the Sun, the good professor backs this up with the illustration in the right-hand margin of two men desperately

Engraving from Camille Flammarion's L'atmosphère: météorologie populaire *(1888). The original caption translates as: 'A medieval missionary tells that he has found the point where heaven and Earth meet …'*

The UNIVERSE after COSMAS 550.

clinging to a speeding globe. 'These men are flying on the globe at the rate of 65,000 miles per hour around the center of the earth (in their minds). Think of that speed!' Anyone wanting to know more was invited to send 25 cents to Ferguson for a book explaining the idea in detail. 'It Knocks the Globe Theory Clean Out,' claims the artwork. 'It will Teach You How to Foretell Eclipses. It is Worth Its Weight in Gold.'

Professor Ferguson was not, of course, the first to see the world as flat; over the millennia the Earth has taken a variety of believed shapes. Homer thought it a disc surrounded by an ocean beneath the dome of the heavens; the Greek philosopher Anaximander (*c.*610 – *c.*546 BC) reckoned it cylindrical; the poet Parmenides (*c.*500 BC) correctly guessed its sphericity; while Anaximenes of Miletus (*c.*585 BC) theorized that it was a disc floating on a cushion of air that occasionally exhaled

A reproduction c.1860 of the universe as a tabernacle, drawn by Cosmas Indicopleustes in his Christian Topography, *AD c.550.*

– these sighs ignited and became the burning stars. Aristotle (384–322 BC) writes of the debate in *On the Heavens*, II, 13.3:

There are similar disputes about the shape of the earth. Some think it is spherical, others that it is flat and drum-shaped. For evidence they bring the fact that, as the sun rises and sets, the part concealed by the earth shows a straight and not a curved edge … Others say the earth rests upon water. This, indeed, is the oldest theory that has been preserved, and is attributed to Thales of Miletus. It was supposed to stay still because it floated like wood and other similar substances, which are so constituted as to rest upon water but not upon air.

He then forms his own conclusion in II, 14: 'Our observations of the stars make it evident … that the earth is circular in shape, but also that it is a sphere of no great size: for otherwise the effect of so slight a change of place would not be quickly apparent.'

Then came Christianity. In the book of Exodus, God spends forty days instructing Moses on the construction of the tabernacle, the portable earthly residence for the divine presence. It was theorized by Christian writers that the description was symbolic, the intricate parts for a harmonious whole being a metaphor for the structure of the universe. Lactantius, a Christian writer from the fourth century, championed this idea and rejected any contradictory theories as paganism; as did the Byzantine geographer Cosmas Indicopleustes in the sixth century, who drew the literal tabernacle universe in this accompanying map. Here the flat *ecumene*, or known world, consisted of a giant mountain surrounded by sea, contained in the curved vault, the walls of which were hidden from our vision by the *stereoma* (heavenly veil).

The Flat Earth idea lived on into the Middle Ages, despite the rational observations of Aristotle and others having established its sphericity so many years before. Contradictors were subject to persecution: Bishop Vergilius of Salzburg in the eighth century, for example, fell foul of Pope Zachary for his disputations – the pontiff decried the spherical Earth theory as 'perverse and sinful'. The Flat Earth theory was eventually forced to retreat into the persistently gloomy corners of pseudoscience, to make occasional reappearances in curiosities, such as that of the unwavering Professor Ferguson.

FONSECA

12°27'N, 54°48'W

Also known as Fonte Seca, Fonesca, Fonzeca, Fonsequa,
San Bemaldo, S. Bernaldo

In 1630, a group of Puritans named the Providence
Company, led by John Pym, drew up plans to establish
a colony far from English shores. Their first choice for
the location of their new home was the island of Fonseca,
which lay to the east of Barbados and was reputedly
a highly fertile land, according to the Spanish. The
revelation soon after that Fonseca could not be found
threw a spanner in those works, however, and instead the
Puritans went on to found Providence Island, 120 miles
(190km) east of Nicaragua in 1631.*

An understandable mistake, for Fonseca had appeared since
1544, when it was drawn by Sebastian Cabot on his world map
as the island of S. Bernaldo, located northeast of Venezuela's
Orinoco delta. Then, in 1589, Hondius entered it on his world
map in the same location as 'Ysla Fonte Seca' (Island of the
Dry Fountain). 'The island of Fonzeca standeth in Degree
of latitude 11¼ ,' states Hakluyt in his *Voyages* published the
same year. In 1628, Charles I prematurely granted the island
to Philip, Earl of Montgomery. The island continued to evade
explorers, but led a full life in rumour: in letters from the
Indies to Spain in 1630 we learn it was reputed to be a
favourite refuge for corsairs, and was considered highly
dangerous for travellers.

Later, on 26 November 1632, Captain Anthony Hilton,
governor of Association Island (Tortuga), requested support for
an expedition he would lead to find the lost island of 'Fonzeca'.
A 40-ton pinnace named the *Elizabeth* was procured for the
voyage, and a group of twenty settlers chosen for the crew, but,
on 26 March 1633, the mission was abandoned in favour of an
expedition to the island of Providence.

Fonseca persisted as an intriguing mystery, and even inspired
two popular, yet wholly fictional, accounts of its discovery.

*Following pages: South of La
Bermuda, and the mythical 'Sept
citez', Hondius drew 'y de fonte
seca' on his world map of 1595.*

---

* *Though the explicit motive was to form a model Puritan colony, the
island's position was considered useful for assaulting Spanish ships, and so
became an iniquitous harbour for privateers. It was attacked and destroyed
by the Spanish ten years later, in 1641.*

The first was *A Discovery of Fonseca* (1682), authored by a 'J. S.', which places the island in the Lesser Antilles and describes it as being inhabited by Amazon women of Welsh origin. The pamphlet describes the habits, customs and religion of the female population, taken 'from the Mouth of a Person cast away on the Place in a Hurricane'. The women are said to be:

a straight handom people, not so black haired as is to be imagined in that hot Climate … they have their houses low and large windows being much delighted with the cool brezes which continually blow there … their garments hanging loose of themselves … their hair hanging in artificial rings shadeing their brests which are round and clear. Their Religion consists in worshipping the Moon …

The second account, *A Voyage to the New Island Fonseca* (1708), purports to be the eye-witness testimonies of two Turkish captains. One of the men, named Aga Sha'ban, states that he landed at Fonseca in 1707 and gives a very different description of the island's population, reporting it to be a predominantly British community of around 16,000, with an African-slave population of about 70,000. The township was a foul and immoral hellhole. Rampant drinking, gambling and promiscuity were encouraged – the sinfulness overwhelming the efforts of the few local priests.

From there, Fonseca began to fade from charts and from memory. By the beginning of the eighteenth century, it had virtually disappeared from maps, until 1866, when, from out of the blue, it pops up in Johnston's *The Royal Atlas*. This occurred despite the rather conclusive previous findings of the United States government brig *Dolphin*, which took a sounding in 1852 at the supposed location and found the water depth to be 2570 fathoms (4700m).

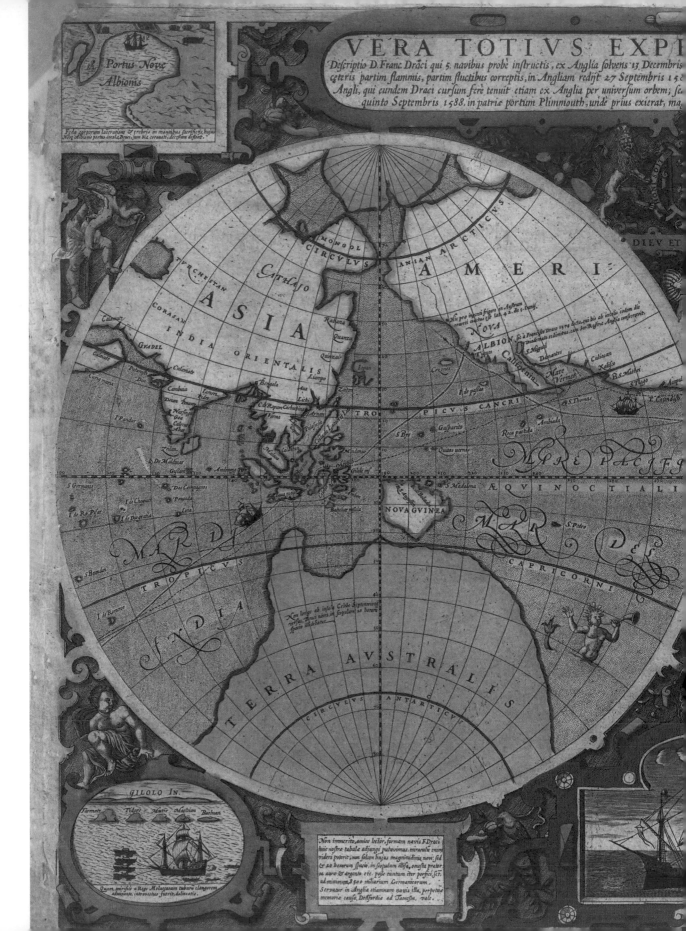

VERA TOTIVS EXPL

Descriptio D. Franc. Draci qui 5. navibus probe instructis, ex Anglia solvens 13. Decembris
ceteris partim flammis, partim fluctibus correptis, in Angliam redijt 27 Septembris 158
Angli, qui eundem Draci cursum fere tenuit etiam ex Anglia per universum orbem; se
quinto Septembris 1588. in patriæ portum Plimmouth, unde prius exierat, ma

DIEV ET

MONGOL

CIRCVLVS ANIAN ARCTICVS AMERI

TVRCHESTAN

CATHAIO ASIA NOVA

CORASAN ALBION

INDIA ORIENTALIS Califormia Mare

GVADEL Rabana Vermeio

Quantu

Quinzai

Calicut Culnabate

TROPICVS CANCRI

Cambaia Lichi MARE PACIFI

Nasinga CaRapan Cachuchina

Goa Verna

Zeilan

De Maldiuar

Gustan Andama AEQVINOCTIALI

Dos Compagnos NOVA GVINEA

MAR DES

MARI CAPRICORNI

TROPICVS

D INDIA

TERRA AVSTRALIS

CIRCVLVS ANTARTICVS

GILOLO In.

Non immerito, amice lector, formam navis F. Draci
huic nostræ tabulæ adjungi putavimus...

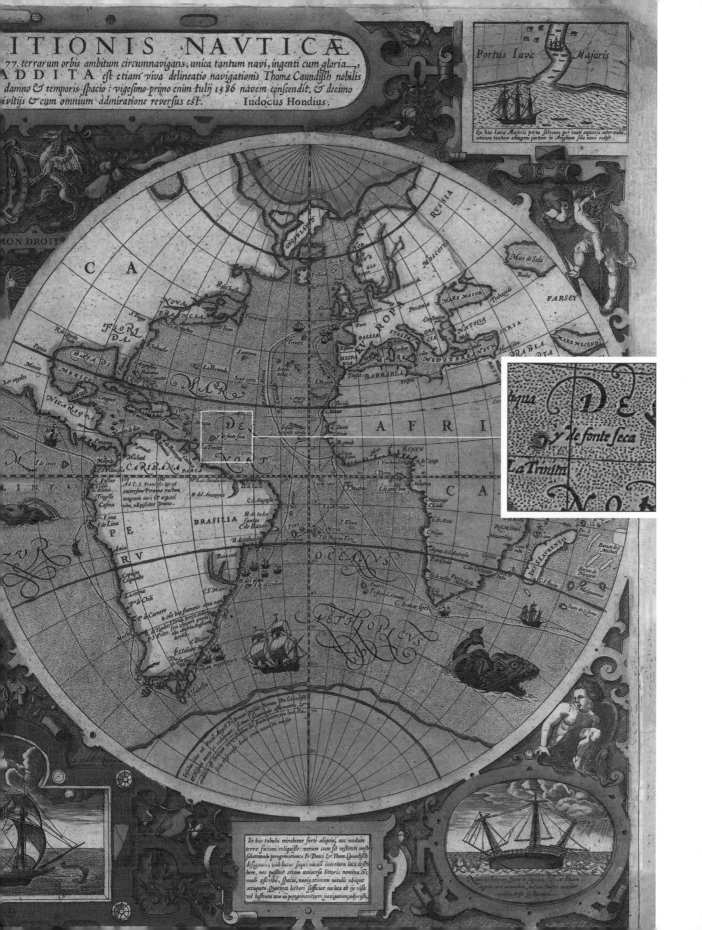

FORMOSA *of George Psalmanazar*

23°30'N, 121°00'E   *Also referred to as Tyowan, Pak-Ando, Gad-Avia*

The Formosa here in question, while geographically identical to the east Asian island known today as Taiwan, is an entirely fictional country. It is the Formosa described in fantastic detail by a Frenchman named George Psalmanazar, who, in the eighteenth century, fooled European society into believing that he was the first Formosan to step foot on their continent.

It was a Jesuit priest who kidnapped him, Psalmanazar said, spiriting him away to Europe and trying to convert him from paganism to Catholicism. Psalmanazar apparently escaped his clutches and, while wandering through Holland, met an Anglican priest named Alexander Innes, who brought him to London and presented him to the bishop of London. He was instantly popular in the city, and enthralled audiences with descriptions of a homeland Europeans knew very little about.

In reality, the impostor Psalmanazar, who took his name from the Old Testament (and whose real name was never discovered), was a blonde-haired, blue-eyed, pale-skinned Frenchman born in the south of France sometime between 1679 and 1684. The spurious tales of his life and travels quickly made him a celebrity in England, and in 1704 his *An Historical and Geographical Description of Formosa* became a publishing sensation, thanks to salacious details of human sacrifice, cannibalism, polygamy, infanticide and other gruesome tales that also played on the strong anti-Catholic and anti-Jesuit attitude of the time. (Praising the Church of England as a 'true Apostolical Church' earned him a loyal following.) The book was wholly fictitious, and borrowed liberally from other contemporary journals, especially accounts of travels to Aztec and Incan sites in the New World, as well as from Bernhardus Varenius's *Descriptio regni Japoniae et Siam* (1649).

The Formosa described by Psalmanazar was a wealthy land, the capital city of which was 'Xternetsa' and whose citizens went about their daily business in the nude, save for a gold or silver plate covering their genitals. Horses and camels were ridden by all, and polygamy was standard. In the event of infidelity, the husband reserved the right to eat his cheating wife – a break from the usual daily diet of serpent-meat.

A portrait of George Psalmanazar from his Memoirs… *(1764).*

Opposite: Formosa, as described in Psalmanazar's An Historical and Geographical Description of Formosa *(1704).*

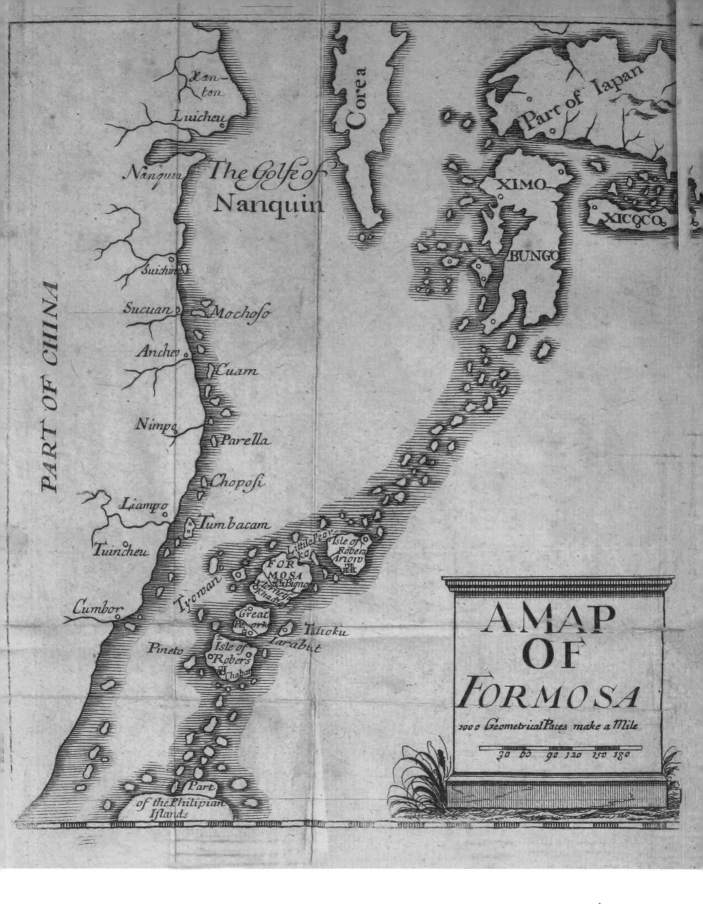

A MAP OF FORMOSA

1000 Geometrical Paces make a Mile

30 60 90 120 150 180

PART OF CHINA

Xanton

Luicheu

Nanquin

The Golfe of Nanquin

Suichin

Sucuan   Mochofo

Ancheo

Cuam

Nimpo

Parella

Chopofi

Liampo

Tumbacam

Tuincheu

Tyovan

Cumbor

Pineto

Little People
ko
FOR
MOSA
St. Bigno
Morilye
Okhabat
Isle of Robert
Ariorv

Great
People
ork
Talioku
Tarabut

Isle of
Robert's
Chabac

Part
of the Philipian
Islands

Corea

Part of Iapan

XIMO

XICOCO

BUNGO

The Funeral, or Way of Burning the Dead Bodies

fig: 4.

A Formosan funeral procession.

Murderers were hung upside and shot full of arrows, while an annual sacrifice was made of the hearts of 18,000 young boys, the bodies of whom were eaten by the Formosan priests. The latter was accompanied by a grisly drawing of 'The Gridiron upon which the hearts of the young Children are burnt'. His sensational description of child sacrifice gave Jonathan Swift the famous trope of cannibalism in *A Modest Proposal*, where his name is mentioned as:

the famous Sallmanaazar, a Native of the Island of Formosa, who … told my friend, that in his Country when any young Person happened to be put to death, the Executioner sold the Carcass to Persons of Quality, as a prime Dainty, and that, in his Time, the Body of a plump Girl of fifteen, who was crucified for attempting to Poison the Emperor, was sold … in Joints from the Gibbet.

As he toured with his Formosan lectures, delighting dinner guests by devouring his meat raw and bloody in Formosan custom, Psalmanazar regularly encountered sceptics, yet survived thanks to ignorance of Formosan customs and his impressive ability to explain away even the most ludicrous of claims. His pale skin? Why, that was because Formosans lived underground. Does the sun shine directly down the chimneys of his country, asked Edmund Halley during a grilling at the

Psalmanazar's invented Formosan alphabet.

Name	Power			Figure			Name
Am	A	a	ao	ỻ	I	I	ᴶﾉ
Mem	M	m̃	m	ᴶ	ᴶ	⅃	ᴶᴸﾉ
Nen	N	ñ	n	ᴜ	ᴜ̆	ᴜ	ᴜᴄﾉ
Taph	T	th	t	ō	Ƃ	Ō	xıÕ
Lamdo	L	ll	l	ſ	F	Γ	ɝ⅃⅃ⅇ
Samdo	S	ch	s	Ƅ	Ɏ	Ƅ	ᴶᴄ⊏ﾉ
Vomera	V	w	u	Δ	Δ	△	ıₒₘᴶⅇ
Bagdo	B	b	b	/	/	/	ᴶᴸⅇⅇ
Hamno	H	ldh	h	ᴸ	ᴸ	ᴸ	ᴶᵤᴸﾉ
Pedlo	P	pp	p	ᴛ̄	ᴛ	Λ	ᴈᴄ⊏ₐ
Kaphi	K	k	x	ᵞ	Ƴ	ᵞ	ₐxıΛ
Onda	O	o	ω	ᴈ	Ɔ	⊐	ᴄₒₗ
Ilda	I	y	i	ₒ	▢	⊟	ᴶⅇⅆﾉ
Xatara	X	xh	x	ᵎ	ᵎ	ᵎ	ıₒₘᴵᴸᴄ
Dam	D	th	d	ᴶ	ᴶ	⅃	ᴶı⁊
Zamphi	Z	tf	z	ᴸ	ᴸ	Ɓ	ₒxᴶıᴸ
Epfi	E	ε	η	ᴄ	ᴄ	Ⅹ	ₐⅾⅈᴄ
Fandem	F	ph	f̃	X	X	✕	ᴈᵤıᴸX
Raw	R	rh	r	ꝑ	ꝑ	▢	Δıᴧ
Gomera	G	g	j	ꝑ	ꝑ	ꝑ	ıₒᵤᴈꝑ

The Formosan Alphabet

pag 122

T. Slater Co.

Royal Society. No, it does not, replied Psalmanazar. It must do! cried Halley triumphantly, for it lies in the Tropics! An excellent point, agreed the Formosan, were it not for the fact that Formosan chimneys are corkscrew-shaped – the sunlight never makes it to the bottom.

He was sent to Christ Church, Oxford, for education, where he stayed for three months, expanding and exaggerating his book for a second edition, including a list of defensive answers to those who challenged the veracity of the first edition, and featuring sensational new details. (It seems from letters from Richard Gwinnet to Elizabeth Thomas sent in 1731 that Psalmanazar claimed to have eaten human flesh himself.)

He eventually confessed all, claiming to have been inspired by a religious experience that convinced him of the sinfulness of his deception. He turned his hand to painting ladies' fans, before studying divinity and spending the rest of his days as an essayist and Grub Street hack, enjoying friendships with prominent English literary figures including Samuel Johnson. It was only when his autobiography *Memoirs of ****, Commonly Known by the Name of George Psalmanazar* was published posthumously in 1764 that the full extent of his masquerade was realized. Johnson was once asked if he had ever confronted his friend, who he held in great esteem, about the deception. 'I should as soon', said Johnson, 'have thought of contradicting a bishop.'

The Formosan Idol of the Devil.

FUSANG

40°33'N, 121°59'W

The story of a 90-year-old missionary monk of AD 499 named Hui Shen (sometimes Huishen, Hwui Shan, and meaning 'Very Intelligent') has been at the core of a hot debate among historians in the last few centuries. The ancient tale, written in the *Liang Shu*, the official history of the Liang dynasty completed in AD 635, takes the form of the Buddhist traveller reporting to the court of Chinese emperor Wu Ti at Jingzhou his discovery of an enormous distant country he named 'Fusang'. The unresolved contention comes from the startling suggestion that this is evidence of the Chinese discovering the New World one thousand years before Christopher Columbus.

Inspired by sailors' stories of a great land on the other side of the 'Eastern Ocean', Hui Shen and a group of monks secured a vessel and headed northeast of Japan to the land of Ta-han (the Siberian Kamchatka peninsula). They then journeyed southeast for a distance of 20,000 *li* (about 6600 miles/10,600km). Along the way the monks interacted with various natives, including owners of domesticated reindeer (Siberians) and men with tattooed bodies, which one could guess to be the Inuit. Finally, the monks reached the 'wonderful land of Fusang.' According to calculations, this could place the monks squarely in the centre of Mexico.

   The soil of the foreign territory was rich in copper, gold and silver, though lacking in iron. The Fusang inhabitants manufactured paper from the bark of a common plant and cloth from its fibres to produce robes or wadding, while wood from the plants went into the construction of their houses. The colour of the people's clothing would change every two years in a ten-year cycle of five colours: blue, red, yellow, white and black. Vegetation was also the principal food source, as well as deer bred for their meat and milk. The Fusang people, who rode horseback, were described as generally law-abiding citizens who were under the reign of a *yiqi* (king) and his team of several officials, with no army or military force to speak of. There were, however, two jails – the one in the north being for those who had committed serious crimes and sentenced to lifelong terms. These prisoners were allowed to get married, and if children were produced the sons would be made slaves and the daughters maids.

Marriage for the citizens was a simple process: if a boy wished to propose to a girl, he would build a cabin beside her home and live there for a year. If she consented, the wedding would go ahead; otherwise, he would be turned away. No form of religion was observed in Fusang, and funerals always took the form of cremation, after a mourning period that varied from seven days for the death of a parent, five days for a grandparent to three days for a sibling. During this time no food or water was to be consumed by the grieving party.

Hui Shen also describes another place, 'Nuguo' (Women's country), which is located 1000 *li* (270 miles/500km) east of Fusang. Here the population consists of beautiful women whose skins are furry, with the hair on their head reaching down to the ground. In the second and third months, they run to the waters and become pregnant, and give birth in the sixth and seventh months. They do not breast-feed their young because they do not have breasts, possessing instead hairs with white roots on the backs of their necks that produce a juice, which they feed to their offspring. At 100 days old, the children can walk, and at four years old reach adulthood. Shen also notes that, when the Nuguo women glimpse men, they run away and hide, and that they chew on a salty grass as an animal would.

There has been much discussion as to which land Hui Shen was describing, or whether the story is complete mythology. It is easy to dismiss the tale as the latter – Yao Silian, who wrote the *Liang Shu*, was using second-hand information from his father's accounts, and no mention seems to have ever been made of Hui Shen outside of this book. Some are certain, however, that Fusang is Mexico, while others suggest that he found himself somewhere in North America, perhaps the Rocky Mountains; Maya is another option, as is Siberia, and the northern islands of Japan. Regardless, word of Fusang found its way to Europe in the eighteenth century, and cartographers produced maps with the land included, placing it alongside other unseen wonders such as the Strait of Anian and the Sea of the West (see relevant entries on pages 12 and 216).

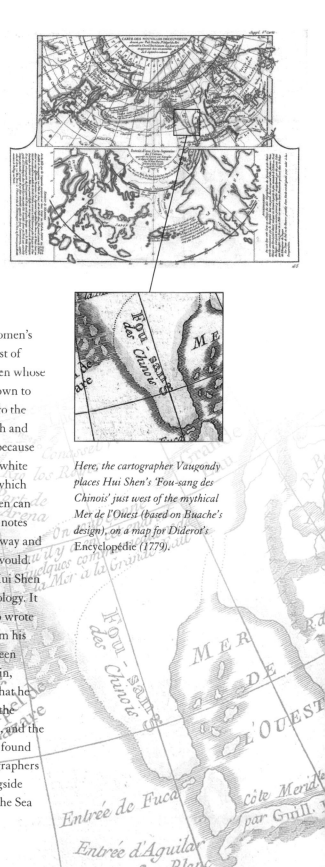

Here, the cartographer Vaugondy places Hui Shen's 'Fou-sang des Chinois' just west of the mythical Mer de l'Ouest (based on Buache's design), on a map for Diderot's Encyclopédie (1779).

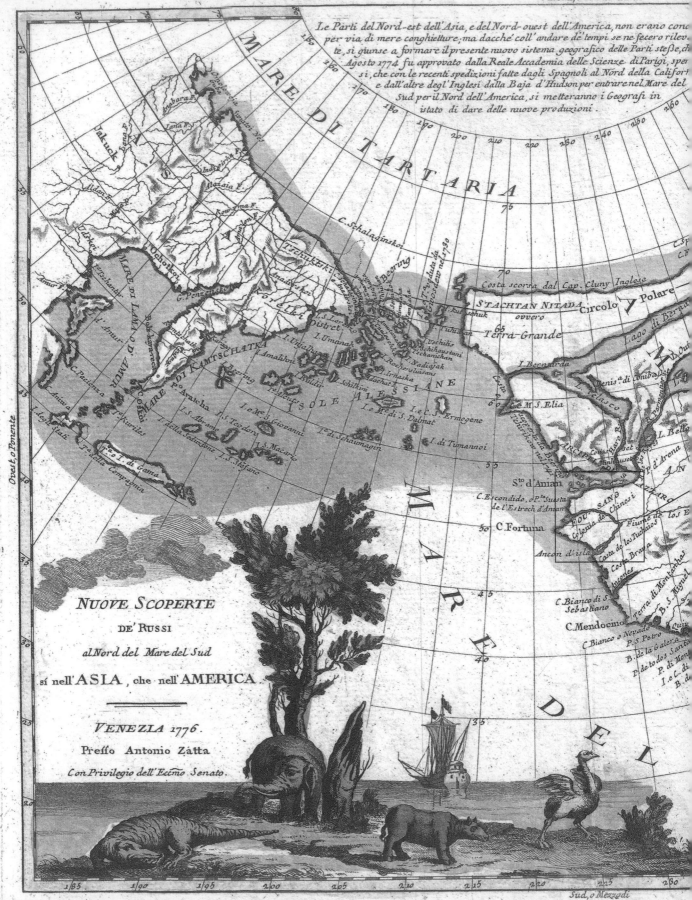

*Antonio Zatta's beautiful
1776 map of northwest
America and northwest
Asia shows Fousang
colonia de Chinesi on
the American west coast,
just south of the equally
nonexistent Strait of Anian
(see Strait of Anian entry
on page 12).*

GAMALAND AND COMPAGNIES LAND

42°12'N, 160°05'E

Also known as Gama Land, De Gama's Land

At the beginning of the sixteenth century, the Malaysian state of Malacca was a powerful port city that controlled the Strait of Malacca, through which passed the shipping between China and India. In 1511, the Portuguese general and conqueror Alfonse de Albuquerque arrived from Goa with an invasion force of some 1200 men, on the orders of King Manuel I of Portugal. After massacring the Muslim natives (but sparing the Chinese, Hindu and Burmese), Albuquerque installed a Portuguese administration with the intention of capitalizing on the trading traffic – Venice would now be forced to buy her spices from Portugal. With the stability of the old regime

Johannes Jansson's Nova et Accurata Iaponiae… *(1658). The 'Landt van Eso' is actually Hokkaido; Compagnies Land is marked to the far right.*

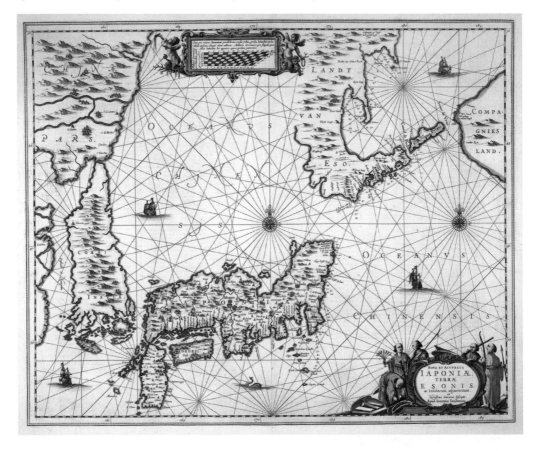

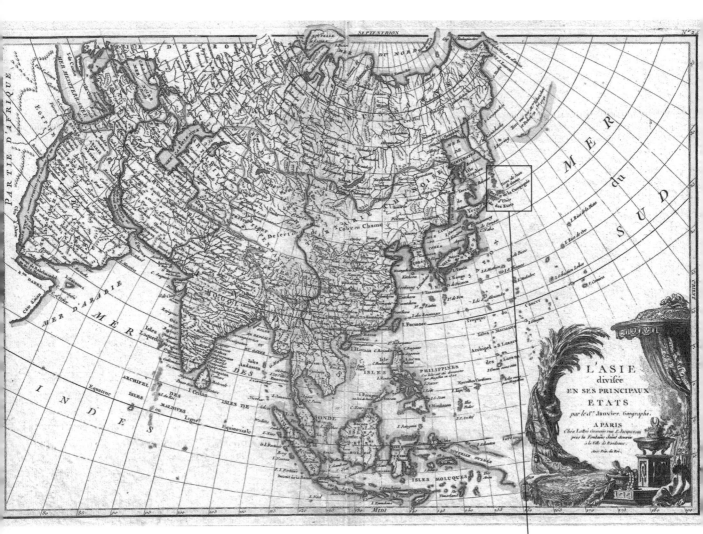

lost, the traders scattered and the Strait of Malacca quickly descended into lawlessness. The corruption extended to the Portuguese administrators themselves, one of whom, a Malacca captain named João da Gama, organized lucrative illegal dealings with the Spanish, trading oriental silks for South American gold. In 1589, da Gama's scheme was rumbled and, to escape sentencing, he fled across the North Pacific to Acapulco. It was during this journey, while north of Japan, that da Gama declared he had sighted land. Despite the dubious character of its identifier, 'Gamaland' was added to early Portuguese maps as a series of small islands and then, curiously, transformed into a much larger theoretical land mass.

Le Sieur Janvier's 1771 map of Asia, showing 'Terre de Jean de Gama' and 'Terre de la Compagnie' east of Yeco (Hokkaido).

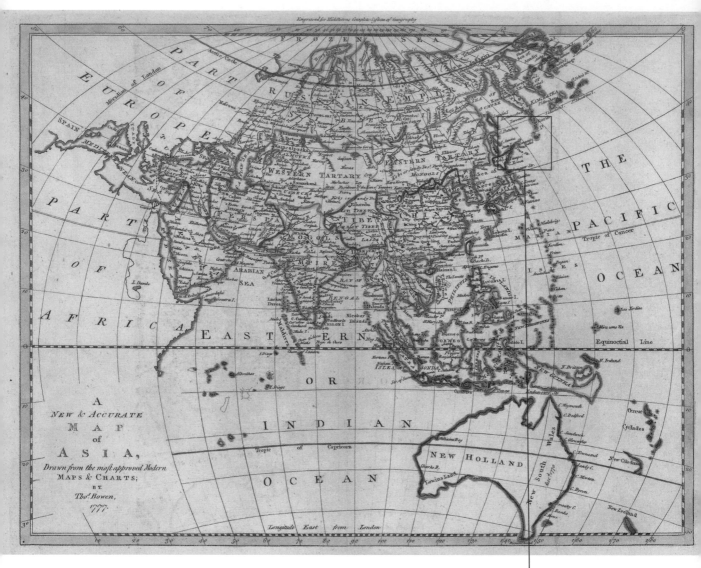

Engraved for Middleton's Complete System of Geography

A
NEW & ACCURATE
MAP
of
ASIA,
Drawn from the most approved Modern
MAPS & CHARTS;
BY
Thos. Bowen,
1777.

The Dutch explorer Matthijs Hendrickszoon Quast searched extensively for Gamaland in 1639, while on a mission to find two other phantoms, Rica de Oro and Rica de Plata, somewhere east of Japan. His fellow countryman Maarten Gerritszoon Vries in 1643 also scoured the Pacific for the land of Gama in 1643, having heard the same rumours of islands rich in gold and silver; but both endeavours failed to find a trace of the country. In the eighteenth century, a sceptical Vitus Bering, a Danish explorer in Russian service, spent three days hunting for the islands, only to have his doubts confirmed. On 16 April 1779, Captain James Cook made a critical note of Gamaland in the third volume of his *Voyage to the Pacific Ocean:*

'Companys Land' shown on
A New & Accurate Map of
Asia, *Thomas Bowen (1777).*

At noon, on the 16th, our latitude was °42 12' [sic] and our longitude 160°5'; and, being near the situation where De Gama is said to have seen a great extent of land, we were glad of an opportunity of contributing to remove the doubts, if any yet remained, respecting this pretended discovery … Mr. Muller relates, that the first account of it was in a chart published by [Portuguese cartographer] Texeira, in 1649; who places it between the latitude of 44° and 45°, and calls it 'land seen by John De Gama, in a voyage from China to New Spain.' Why the French geographers have removed it five degrees to the eastward, we cannot comprehend; unless we suppose it to have been to make room for another fresh discovery made by the Dutch, called Company's Land.

This last mention of another fiction, the neighbouring Company's Land (also known as 'Compagnies Land' and supposedly located at 45°56'N, 150°02'E), was created in the summer of 1643 by Vries as he hunted in vain for Gamaland. He sailed his ship, the *Castricum*, between two new islands, and named the one to the south 'Staten Landt' after the Dutch States General, and the one to the north 'Companijs Landt' in honour of the Dutch East Indies Company. In fact, the islands already had a name – the Dutchman having passed between Iturup and Urup – of the volcanic Kuril Islands in the Sea of Okhotsk. When details of these were passed back to European cartographers, 'Companijs Landt' was seized on as corroboration of Gamaland, with which it was conflated. Its size was exaggerated accordingly, and it haunted European maps for more than a century.

GREAT IRELAND

58°15'N, 34°19'W   *Also known as Hvítramannal, Hvítramannaland, Írland hið mikla,*
White Men's Land. Hibernia Major, Albania

Few details survive about Ari Marsson, the Icelandic
explorer of the late tenth century, but the story of his
legendary voyage of 982 or 983 in which he discovered
a new country is recorded in some detail in Chapter 43
of the *Landnámabók* (Book of Settlements), a medieval
Icelandic history of the Norse colonization in the ninth
and tenth centuries:

[He] was driven by a tempest to Hvítramannaland [White
Men's Land] which some call Írland hið mikla [Great Ireland or
Hibernia Major] which lies in the western ocean, near to Vínlandi
hinu góða [Vinland the Good], six days' sailing west from Ireland.
Ari couldn't get away, and was baptized there. This story was first
told by Hrafn Limerick-Farer who spent a long time at Limerick
in Ireland. Thorkel Gellisson quoted some Icelanders who had
heard Earl Thorfinn of Orkney say that Ari had been recognized
in White Man's Land, and couldn't get away from there, but was
thought very highly of.

Further information as to this strange place of Great Ireland
was mentioned in another saga, which states:

To the south of habitable Greenland there are uninhabited and wild
tracts, and enormous icebergs. The country of the Skraelings lies
beyond these; Markland beyond this, and Vinland the Good beyond
the last. Next to this, and somewhere beyond it, lies Albania, that
is, Hvítramannaland, whither, formerly, vessels came from Ireland.
There, several Irishmen and Icelanders saw and recognized Ari, the
son of Mar and Kotlu, of Reykjanes, concerning whom nothing had
been heard for a long time, and who had been made their chief by
the inhabitants of the land.

In the saga of Eric the Red, a captured Skraelinger (a
Norse term for the indigenous Greenlanders) describes the
inhabitants of Great Ireland as 'dressed in white garments,
uttered loud cries, bore long poles, and wore fringes'. Snatches
of Great Ireland references crop up in other later sources as
well. The twelfth-century Arab geographer Muhammad

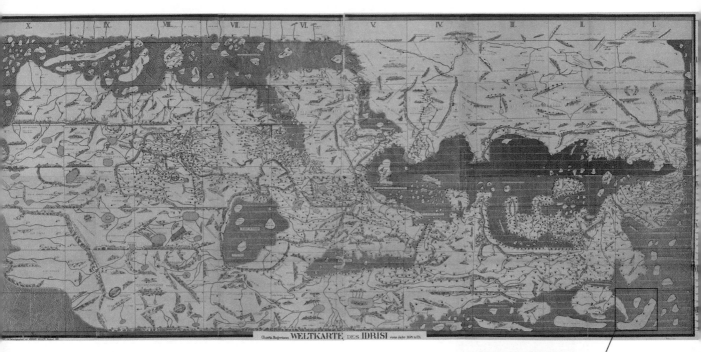

Charte Rogerianum WELTKARTE DES IDRISI vom Jahr 1154 n.Ch.

al-Idrisi mentions 'Irlandah-al-Kabīrah' in his *Tabula Rogeriana*, stating that 'from the extremity of Iceland to that of Great Ireland' the sailing time was 'one day'. Elsewhere, there is a brief mention of the people of Great Ireland in the *Hauksbók* (Book of Haukr), a medieval Norse manuscript that describes the island's inhabitants as *albani*, possessing white hair and skin.

Great Ireland has never been satisfactorily identified. The name originates from reports that the native tongue sounds similar to that of the Irish, as well as stories of its origination through colonization by Irish monks, linking it with the stories of St Brendan's Island and other Irish immrama (see St Brendan's Island entry on page 202). In 1888, the Norwegian historian Gustav Storm concluded that the story of Great Ireland was pure legend; and yet others have speculated as to the geography of Marsson's landfall with wildly varying theories. Marion Mulhall, in 1909, decisively identified the coast of Florida as the ground on which the Icelander trod, while Gustavo Nelin, in *La saga de Votan: contactos vikingos en México en el siglo X* (1989), goes a step further with an especially creative theory: that Marsson made his way to Mexico, where he made such an impression on the natives with his strange appearance and habits that he became the Mesoamerican feathered serpent deity Quetzalcoatl.

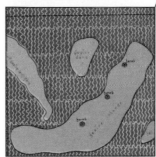

A later copy of the world map from the Nuzhat al-m ushtāq fi'khtirāq al-āfāq *(The Book of Pleasant Journeys into Faraway Lands) informally known as the* Tabula Rogeriana*, completed by Muhammad al-Idrisi in 1154. Great Ireland is here marked 'Gezire Irlanda' and can be found in the bottom-right corner.*

📍 GREAT RIVER OF THE WEST

49°09′N, 113°11′W   *Also known as: Buenaventura River, Río Buenaventura, Rivière longue, La belle rivière, fleuve de L'Ouest, The Long River*

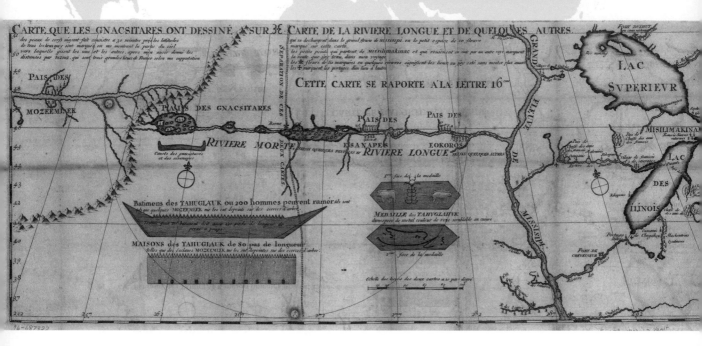

'Thank God, I am now return'd from my Voyage upon the Long River, which falls into the River of Missisipi', writes Baron Lahontan in Letter XVI of his *New Voyages* (1703). 'I would willingly have trac'd it up to its Source, if several Obstacles had not stood in my way.' The imaginary river of Louis-Armand de Lom d'Arce de Lahontan, Baron de Lahontan, is today a dust-caked historical curiosity filed deep in the back of the cabinet; but, at the time of publication, the puffed-up travel journal was phenomenally popular, with thirteen editions printed in fourteen years, and it had a powerful influence on the contemporary perception of North American geography.

Born in France in 1666, by the age of seventeen the aristocrat was a lieutenant in the *troupe de la marine*, the arm of the French Army serving in Canada, during which time he ventured further south on missions to Fort Michilimackinac in northern Michigan, and later to Fort St Joseph near Detroit.

Map of the Rivière longue from Nouveaux Voyages de M. le Baron de Lahontan dans l'Amerique Septentrionale *(1703).*

All the while, the young soldier was drafting letters home to an elderly patron, recording his experiences and the data he was gathering about the new land, drawn from both his own investigations and local native guides. These documents form the basis of his book, and in Letter XVI, as he explores the great northwest in 1688–89, he makes his most outrageous claim: of coming across a giant *Rivière longue* that ran from the Rocky Mountain region eastwards across central North America, to empty in the Upper Mississippi. It was a report that would have been welcomed at the time, for those hunting a northwest passage were especially keen on the possibility of a transcontinental waterway running through temperate country. (Captain George Vancouver's comprehensive survey of the northwest coast with two Royal Navy ships in 1794 would puncture this dream.)

Bellin's map of North America (1743) showing the 'Riviere de l'ouest'.

Pag. 15. Tom. III.

As he sailed along his *Rivière longue*, Lahontan makes observations about local faunae, taking the time to refute the popular medieval belief that beavers, when hunted, would gnaw off their own testicles and hurl them at their pursuer as a distraction. He then goes on to provide a thoroughly dubious description of his dealings with a tribe of natives called the 'Essanapes' and their king, who was carried everywhere by six slaves and had leaves scattered in his path. Upriver, Lahontan then encounters the 'Gnacsitares', who tell him that the *Rivière longue* originated in the tall mountains to the west, the other side of which flows another river down to a great salt lake 300 leagues (1670km) in circumference. Around this lake can be found the 'Tuhuglauk' nation, bearded natives with pointed caps in dress down to their knees, who live there in six cities of stone and a hundred towns. (Alas, Lahontan was unable to convince any of the Tuhuglauks to accompany him back to Canada as proof, for they cared 'but little for riches'.) Lahontan then moved on to the Mississippi, and his account ends with a pitch for the job of leading a well-funded expedition further into native territory: 'I should esteem it my happiness to be employed upon such an enterprise both for the glory of His Majesty, and my own satisfaction.'

Baron Lahontan gathered his information from natives throughout his expedition. Here he is showing local natives a painting of the sky with the sun, moon and stars (Voyages … dans L'Amerique Septentrionale, 1703).

Most cartographers of the eighteenth century, including Herman Moll, Henry Popple, Guillaume de l'Isle and John Senex, seized on this first-hand information and faithfully reproduced it. On his *Carte du Canada*, de l'Isle drew the great salt lake of Lahontan's Tuhuglauks (see Sea of the West entry on page 216 for more on this particular idea) as well as the *Rivière longue* in precise detail. Moll included the river on his map of the world used in several works in 1712, and as late as 1765 Samuel Engel published a map making full use of Lahontan's mistaken details.

Could Lahontan have encountered an existing river and simply mistaken its identity? If so, it is difficult to imagine which one – for the size, length and unswerving course of the *Rivière longue* would seem to rule out the Minnesota; and the Missouri joins the Mississippi much further south. The latter was the explanation offered by reviewers in 1816, despite the fact that Lahontan specifically mentions making a separate exploration of the Missouri. The consensus is that his errors were the result of a heavy and blind reliance on details passed to him by locals, as well as a disposition for fashioning his own idea of the facts for added excitement, although his tendency to fib was milder than most. As Francis Parkman put it in 1877, Lahontan was 'a man in advance of his time … He usually told the truth, when he had no motive to do otherwise, and yet was capable at times of prodigious mendacity.'

Forty years after Lahontan's journey down the mythical river, the French explorer Pierre Gaultier de la Varennes et de la Vérendrye and his sons were scouring the Assiniboine River in Manitoba, Canada, for a River of the West drawn on a map for them by a Cree native named Auchagach, that showed a connected chain of lakes and rivers from Lake Superior to a sea in the west. La Vérendrye never found this route, but drew the sea on his maps, labelling it 'Mer Inconnue'. No account was written of this journey, but in 1913 a lead plate was discovered buried in the earth in South Dakota, placed there by the sons of la Vérendrye, claiming the territory for Louis XV.

GROCLANT

81°07′N, 74°23′W

No one is entirely certain as to how a country covered almost entirely by a thick sheet of Arctic ice came to be called 'Greenland', but the prevailing theory is that it was an early exercise in false advertising. In *c*.982, an exiled Erik Thorvaldsson (Erik the Red) led a group of Icelandic settlers in fourteen ships across the seas to the northwest. Upon finding the great island, they established three colonies on the southwest tip, and named their new home 'Groenland', supposedly in the hope that the verdancy would entice other Icelanders to join them.

One of the earliest known maps to show Greenland was drawn in 1427 by the Danish geographer Claudius Clavus, and shows the country joined to the northernmost part of Europe. Clavus was an influential authority, and this was how Greenland was perceived for years. In 1467, Clavus produced an updated design, but again it shows Greenland as part of mainland Europe. The country then has a rather peripatetic cartographic life: the *Catalan Map* of *c*.1480 has a rectangular 'Illa Verde' parallel to Ireland, in the position of the mythical Hy Brasil (see relevant entry on page 130) also further south; an anonymous map produced at approximately the same time puts it in roughly the correct position; the Martin Behaim Globe of 1492 transforms it into an Arctic peninsula above Norway; and Juan de la Cosa plants it as part of a scatter of islands to the north of Iceland. Amid all this confusion, one particular error emerged from the jumble of names, shapes and positions: Greenland was split into two. An entirely new island, named 'Groclant', began to be drawn alongside 'Groenland' on maps.

    The aforementioned anonymous map of *c*.1480 shows both 'Gronland' and 'Engroneland'; but the most colourful examples occur in the sixteenth century, when Greenland had become sought after (and fought over) by the Portuguese and the Danish. 'Groenland' would be translated by cartographers into their own language, and eventually the two islands came to co-exist as separate entities. Many maps made in this century, and indeed into the seventeenth century, show the original Greenland alongside 'Green Island' variations such as Isla Verde, or Insula

*On Ortelius's 1570
Septentrionalium regionum
(northern region) map, Groclant
is drawn in the top-left corner.*

Viridis. Mercator was the first to show 'Groclant' in his 1569 map
of the polar basin (see Rupes Nigra entry on page 200). The map
featured here is a study of the 'northern region' from 1570, taken
from Ortelius's atlas *Theatrum Orbis Terrarum*. Here one can
clearly see the land of Groenlandt (its 'lushness' emphasized by
the colourist), and then, to the northwest, its nonexistent
neighbour Groclandt. The miniaturization of Greenland's
836,109-sq. mile (2,165,512-sq. km) area is also a striking point
of curiosity on this map, as is the enormous presentation of the
mythical 'Estotilant' (also spelled Estotiland) in the northwest
corner of the map (see Phantom Lands of the *Zeno Map* entry
on page 240 for more on this). In the southwest corner, we also
find the phantom isles of St Brendan and Brasil and, to their
north, Friesland (see St Brendan's Island entry on page 202 and
Hy Brasil entry on page 130).

'Groclandt' would seem to have emerged from confusion
over the spelling of Groēland – one accidental smudging of
the 'ē' and without a tilde it could well appear to be a 'c'. The
mistake would have been compounded by awareness of the
existent Baffin Island, located to the west of Greenland. A
simple error, but one reproduced by authors such as Michael
Lok on his map published by Hakluyt in 1582, in which he
labels the island 'Jac. Scolvus Groctland', possibly after a lost
report of discovery; and by Mathias Quadus on a map of 1608.
By 1610, the area was the subject of searching by Sir Martin
Frobisher, John Davis and others; but, when all failed to turn
up a trace, Groclant was dropped from further mapping,

HY BRASIL

51°N, 17°34'W

Also known as Hy-Breasal, O Brasil, Brazil, Brasile, Bracie, Bresily, Bersil, Brazir, Braziliae, Bresiliji, Branzilae, O'Brassil, Insula Fortunatae (Fortunate Island), the Isle of the Blessed

On the ocean that hollows the rocks where ye dwell,
A shadowy land has appeared, as they tell;
Men thought it a region of sunshine and rest,
And they called it Hy-Brasail, the isle of the blest.

Hy Brasail – The Isle of the Blest, *by Gerald Griffin*

An imaginary island in the North Atlantic, sometimes located off the coast of Ireland but also as far south as the Azores, Hy Brasil is frequently depicted as a circular land through which runs a strait or river. The origin of the name is something of a mystery, though some scholars link it to the northeastern Irish tribe Ui Breasail. (It has nothing to do with the South American country of the similar name, the etymology of which

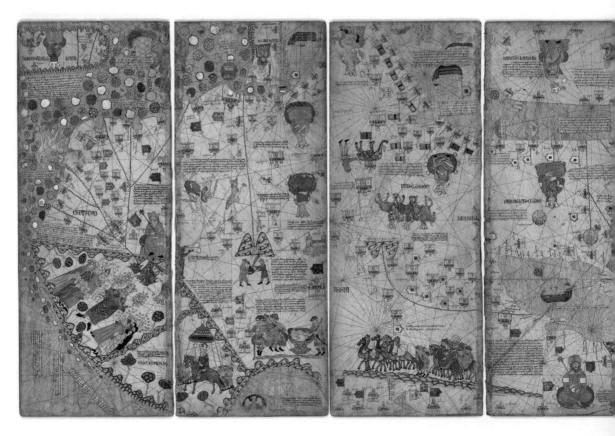

is to do with the brazilwood tree, which, in Portuguese, is *pau-brazil*, meaning red like an ember.)

In the sea tales of Celtic folklore, the immrama, Hy Brasil bears similarities to the myth of Atlantis (see relevant entry on page 24). It was said to be a wondrous paradise, offering eternal happiness and immortality to its inhabitants. The island was ruled over by King Breasal, the high king of the world, and emerged from the depths of the Atlantic every seven years, when the king would hold court for a short period, and then the entire island would disappear back under the water.

Hy Brasil was inscribed on the portolan map of Angelino Dalorto *c.*1325, labelled as 'Insula de monotonis', and later modified in 1339 to 'Insula de Brazil'. Other cartographers then included it in their own work, using it to fit conveniently with various rumours of islands in the area. From here, the island of Brasil would, astonishingly, continue to be included on maps until well into the nineteenth century, thanks to occasional reported glimpses and the seductiveness of a compelling mythology. In contrast to the gradual degradation of the typical phantom because of increased naval traffic over

'Insula de Brazil' shown on the Catalan Atlas (1375), created by Abraham Cresques of Majorca, and commissioned by Charles V of France. The enormous atlas (here presented south-to-north) gives a complete picture of the geographical knowledge of the Middle Ages.

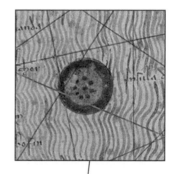

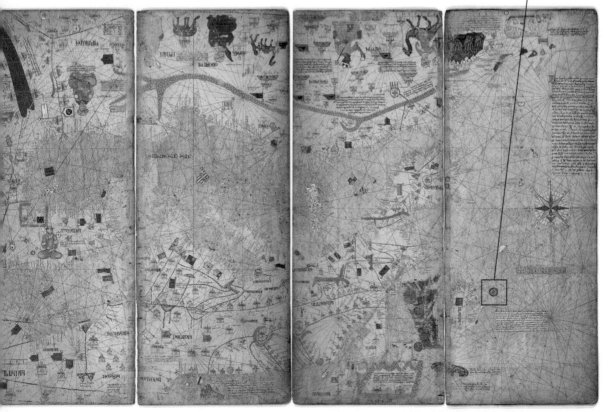

time, the curious thing about Hy Brasil is that official notations of its location become *more* specific.

Jeffrey's *American Atlas* (1776) clearly places the island of 'O'Brazil' at 51°N, 17°34'W. In 1807, the British Admiralty's Hydrographic Office charted Hy Brasil at the coordinates 51°10'N, 16°W, despite the fact that the area was now frequently sailed and no reports of the island had ever been filed. In John Purdy's *Chart of the Atlantic* (1832) one finds it recorded at 50°50'N, 15°20'W, though it is cautiously downgraded to 'Brasil rock'. Finally, in 1865, it was purged from the map by the English geographer Alexander G. Findlay.

Several fact-finding voyages to Hy Brasil were launched over the years, starting in the fifteenth century with William of Worcester's 80-ton vessel captained by Thomas Lloyd in 1480, which was forced to turn back by severe storms almost immediately. The next year, the *Trinity* and the *George* returned empty-handed. In 1498, Pedro de Ayala, the Spanish envoy to London, reported to King Ferdinand and Queen Isabella that the English had sent several missions annually for the past seven years in search of the island. In 1633, Captain David Alexander was hired by Lord Lorne to complete a survey and comprehensive report of Hy Brasil, which also was not achieved.

Though these expeditions always returned fruitless, the obsession with finding Hy Brasil never abated and, in 1675, interest was stoked further by the publishing of a satirical pamphlet by Richard Head. Purporting to be a letter from 'William Hamilton of Londonderry', writing to a cousin in London, the pamphlet was titled *O-Brazile, or The Inchanted Island: Being a perfect relation of the late discovery and wonderful dis-inchantment of an island on the north of Ireland: With an account of the riches and commodities thereof*. It provides a vivid insight into how the island was imagined to be at the time.

'Brazil' marked on Sebastian Münster's 1628 map of Europe.

Written in the style of Henry Neville's 1668 utopian work *Isle of Pines* (which Head derides as a 'monstrous Fiction'), the letter recounts the experience of a sea captain named John Nisbet, who stumbled across the 'Island of Brasil' amid a terrible thick fog. An armed landing party of four men entered a dense forest, which after less than a mile gave

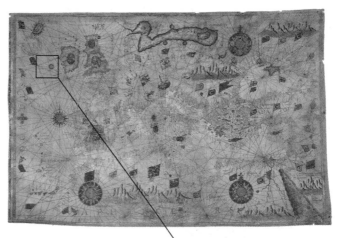

way to a beautiful green valley 'wherein were many Cattel, Horses and Sheep feeding'. Glimpsing in the distance the towers of a castle, the men headed in its direction, hoping to find inhabitants, but found it to be deserted. After hearing 'a most terrible, hideous noise' in the night, the men awoke the next morning to find on the shore 'a very antient grave Gentleman, and ten Men following him bare-headed (as if his servants)'. The man called to the master 'in the old Scotch Language' and invited the crew to join them, assuring no harm would come to them. When they landed, the gentleman told them they were 'the most happy sight that Island had seen in some hundred years; that the Island was called *O Brazile*.' He explained that he and several others had been locked away in the castle by a tyrannical necromancer, 'and the whole Island, a Receptacle of Furies, made (to Mortals) unserviceable, and invisible, until now, that the cursed time was expired'.

Brazil on Giorgio Callapoda's Carte de la Mer Méditerranée *… (1565), kept in the National Library of France.*

The crew was then thrown a feast by the islanders and returned to their ship, soon arriving back in Killybegs, Ireland, to tell their story. Hamilton wrote about several further voyages launched to O Brazile, including one made recently by several godly ministers. 'But at the writing hereof', he notes, 'I heard nothing of their return.' He assures his cousin the story is true, for 'beside the general discourse of the Gentlemen of the Countrey, I had it from Captain Nisbet his own mouth.'

Whether it was due to Head's pamphlet or faith in the folklore, there were many believers in Hy Brasil. This included the eminent (and eminently pragmatic) Robert Hooke, who early in 1675, according to his diary, met with Francis Lodwick at Garraway's Coffee House in London, where the two natural philosophers discussed 'O.Brazill and longitude'.

JAVA LA GRANDE

12°01'S, 124°01'E

From the time of Aristotle, Europeans believed that the southern hemisphere was dominated by a vast continent, commonly referred to as Terra Australis Incognita (unknown land of the South). The logic of the hypothesis was based on a simple principle – to counterbalance the weight of the known northern lands and maintain the world's stability, there must be a southern land mass of comparable weight. In 1520, the first evidence to support the idea appeared to have been found: during his round-the-world expedition, Ferdinand Magellan discovered what he presumed to be a vast nation just opposite the very southern tip of South America. What he actually saw was the archipelago of the Tierra del Fuego, but after studying his report it was later speculated by the Spanish sailor Francisco de Hoces that Magellan had discovered the northern coast of the theoretical Terra Australis. (This particular misbelief is explored in full in the Terra Australis entry on page 224.)

The European discovery of Australia was, at first, also thought to be proof positive of this counterweight continent. As it is most commonly told, the history begins in 1606, when the first significant charting of Australia was made by the Dutch ship *Duyfken*, skippered by Dutchman Willem Janszoon. In 1629, two Dutch sailors then inadvertently became the first European immigrants to Australia when their ship, *Batavia*, collided with the coral of the Houtman Abrolhos Islands, about 25 miles (46km) off western Australia. After sparking a failed mutiny, they were abandoned on the mainland. In 1644, the Dutch sailor Abel Tasman gave the great southern land the name of New Holland; then, in 1770, Captain Cook landed at Botany Bay and claimed the entire east coast of Australia, naming it New South Wales. In 1804, Matthew Flinders took the name 'Australia' from the 'Terra Australis' myth and proposed it as a suitable alternative to New Holland (while also concluding that there was 'no probability' of finding any other land further south).

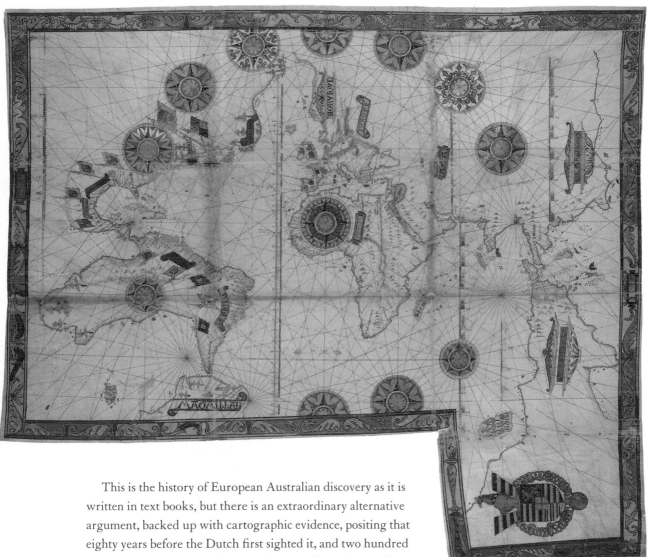

This is the history of European Australian discovery as it is
written in text books, but there is an extraordinary alternative
argument, backed up with cartographic evidence, positing that
eighty years before the Dutch first sighted it, and two hundred
years before Captain Cook stepped onto its shores, it was the
Portuguese who were the first Europeans to discover Australia.
The evidence for this dates back to Marco Polo: in Book III
of his *Travels* he journeys from China to India, via Champa,
Locach and Java Minor (Sumatra) and past 'Java Grande'
(which he never visited). 'Departing from Ziamba', he wrote,
'and steering between south and southeast, fifteen hundred
miles, you reach an island of great size, named Java, which
according to the reports of some well-informed navigators, is
the largest in the world, being in circuit above three thousand

*An enormous world map by
Guillaume Brouscon, one of the
famous Dieppe cartographers,
showing a large Terra Java as
part of the Austral land (see Terra
Australis entry on page 224).*

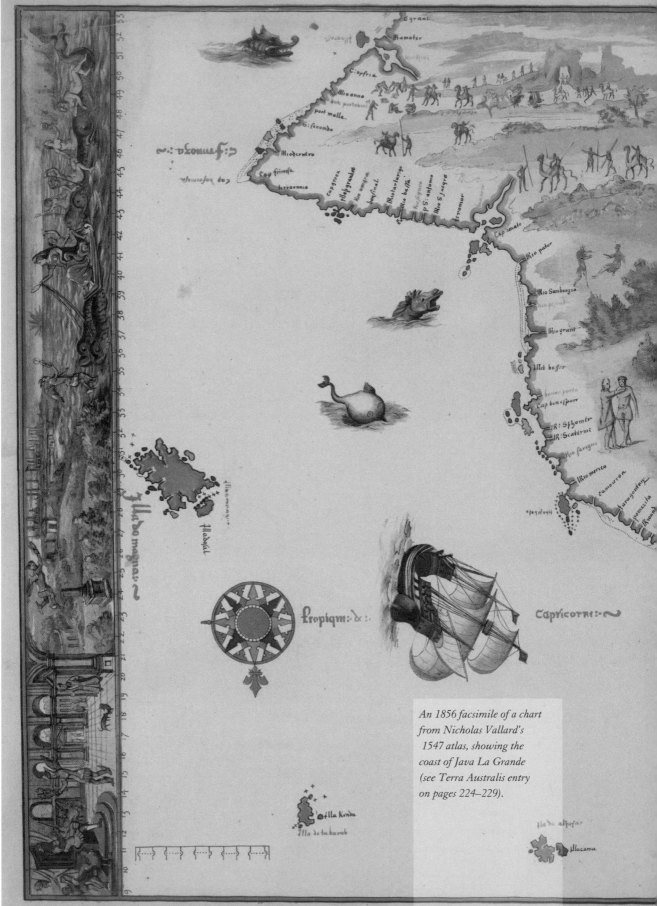

An 1856 facsimile of a chart from Nicholas Vallard's 1547 atlas, showing the coast of Java La Grande (see Terra Australis entry on pages 224–229).

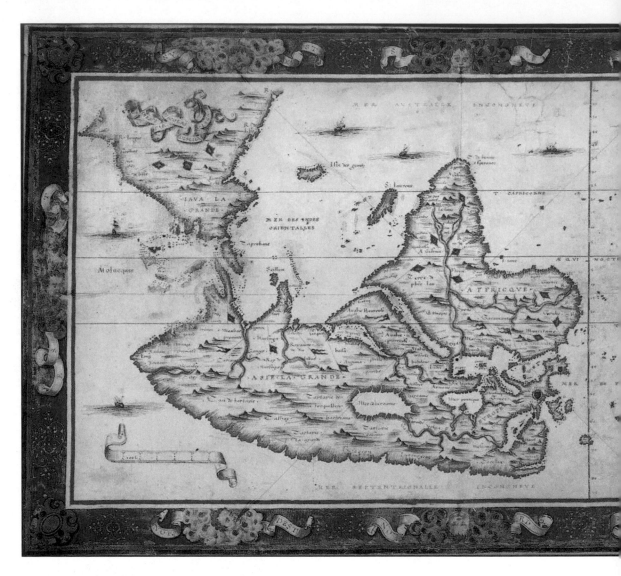

miles. It is under the dominion of one king only, nor do
the inhabitants pay tribute to any other power. They are
worshippers of idols.' Because of some accidental substituting
of place names in the 1532 edition of Polo's account, his
geography became greatly confused, and after Magellan's
expedition Java Minor was placed north of Java Grande (the
various editions of his book differ wildly in such details). An
enormous island was created, and attached by the mapmakers
at Dieppe to Terra Australis, as we can see on Guillaume
Brouscon's 1543 world map (on page 135). This notion seemed
to be confirmed by the account of Ludovico di Varthema, an
Italian who visited Java in 1505 and wrote that it 'extends
almost beyond measure'.

The world drawn by Nicolas Desliens in 1566, showing an enormous Java la Grande on the left side.

Another interesting example of this sixteenth-century perception is demonstrated by the map on pages 137–8, a nineteenth-century reproduction of Nicholas Vallard's 1547 chart produced at Dieppe, which was grandly claimed by its nineteenth-century owner Sir Thomas Philips to be the first recorded map of Australia (probably to boost the importance of his library).

The Dieppe cartographers were the undisputed authority on contemporary geography, and so the observation that some of the maps they produced in 1540–70 feature Portuguese flags over Java la Grande, a giant land in roughly the same location as Australia, has led some to argue that the mapmakers were privy to information that the Portuguese trading empire extended as far as Australia. The subsequent disappearance of this knowledge, maintain the theorists, could be explained by the devastating earthquake at Lisbon in 1755, in which records would have been destroyed. The issue remains something of an enigma.

JUAN DE LISBOA

26°17's, 52°48'E

The curious case here, as drawn by Pieter Goos in 1680, takes the form of islands Juan de Lisboa and Dos Romeiros ('two pilgrims'). Juan de Lisboa is placed a few hundred miles or so from the eastern Madagascan coast, while Dos Romeiros is located further east. The islands are then shuffled around like a game of three-card Monte by later cartographers and given various new coordinates, until finally, in 1727, the two are combined as one by the French cartographer Jean Baptiste D'Anville and placed south of the isle of Bourbon (now known as La Réunion), off the east coast of Madagascar. No one really knew what to do with Juan de Lisboa – details were scarce for years (though not scarce enough to warrant erasure), until rumours of it being a secret buccaneer hideout began to emerge in the second half of the eighteenth century.

Conrad Malte-Brun's *Universal Geography* (1827) tells us that, in 1770, on the Isle of France (Mauritius), notes from incoherent and contradictory journals of travellers were circulated as a matter of general interest, and some of these were included in a memoir on the Isle of Bourbon. This was presented to the general committee of the India Company in 1771, and established 'that the island of Juan de Lisboa appeared imaginary to those navigators only who had not found it out'. As verification of its existence, the paper went on to claim 'that a bucanier had disembarked on it, *not more than six years ago*, and had killed, according to his own account, twelve or fifteen oxen in less than two hours!' The testimony of a M. Boynot is then quoted,

On Pieter Goos's Nieuwe Pascaert van Oost Indien *(1680), published by van Keulen, 'I. de Juan do Lisboa' lies just to the east of the southern tip of Madagascar.*

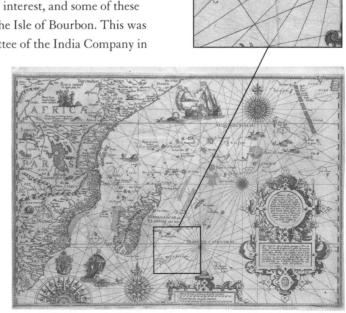

who 'assures us that he had seen and sailed round it towards the end of the year 1707, in returning from the Isle of Bourbon to Pondicherry.' How could one doubt the veracity of this gentlemen, wonders Malte-Brun sarcastically, when he also asserts that: 'he is indebted for this discovery to some bucaniers at that time on board his ship, and takes care to tell us that, by passing to the south of Madagascar, he very much shortened his passage'. (This last claim is most dubious, for the winds and currents would have been against him.)

All was then quiet on the Juan de Lisboa front until 1772, when a Captain Sornin, passing from the Cape of Good Hope to the Isle of France, encountered cataclysmic storms ('the sea very high, the air much heated'), during which he sighted the southern tip of Madagascar, before putting in at Rodrigues island, an outer island of the Republic of Mauritius. He then measured himself 3 leagues (16.5km) short of Rodrigues' position, and so concluded this landfall to be on a new island. Such confusion in stormy weather is enough to leave us doubtful, but the authorities of Isle of France ordered missions to confirm the position of Juan de Lisboa. Records show that a M. de St Felix commanded a search in 1773, as did M. Corval de Grenville in 1782 and 1783; but these turned up empty. The French traveller, a M. Rochon, adds in his journal of his voyage to the East Indies at the time that: 'In returning from Madagascar, we thought at one time that we perceived the island of St Juan de Lisboa, but the illusion was caused by clouds, to which the most experienced mariners are too often exposed.'

This would seem to explain the mysterious elusiveness of the island, but then a M. Epidariste Collin of Isle of France claimed to have been assured by the secretary of the government of Mozambique that their archives held the instructions for the evacuation of the 'Portuguese colony of Juan de Lisboa', as well as an inventory of belongings transported from the island to the coast of Africa. Collin, however, had been unable to confirm the existence of this document with his own eyes.

There is no such island as Juan de Lisboa and, considering the frequency of its sighting in circumstances of wild storms and visually deceptive conditions, it seems likely that it is the result of confusion with the Madagascan coast, Réunion Island or even, perhaps, Mauritius, as M. Epidariste Collin had himself theorized, before learning of the strange story of the island's supposed evacuation to Africa.

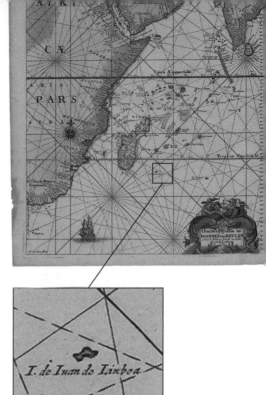

Langren's Delineatio orarum maritimarum *(1596).*

LOST CITY OF
THE KALAHARI

23°55'S, 21°53'E

The first thing you should know about William Leonard
Hunt is that he was a talented inventor. The Victorian
showman, known as 'the Great Farini', devised a
theatrical stage device for propelling a person into the air,
which he adapted to create an early (if not the earliest)
human cannonball show. He debuted this machine at
London's Royal Aquarium on 2 April 1877, when he shot
a 14-year-old girl named Rosa Richter 70ft (21m) across
the auditorium into a net.

Hunt's gift for innovation was also evident in his florid sales
hyperbole, which rivalled the ingenuity of that king of humbug
P. T. Barnum. In fact, Barnum signed up Hunt and Rosa (who
performed as 'Zazel') in 1880, and the cannonball duo toured

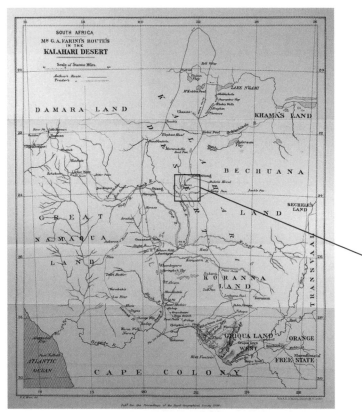

The map that William Leonard
Hunt presented to the Royal
Geographical Society as part of
a report on his exploration of
the southern African region: A
Recent Journey in the Kalahari
by G. A. Farini *(1886). The
indication of 'Ruins' refers to his
claim to have discovered a buried
city of an ancient civilization.*

across America with the world's greatest circus to tremendous success. But Hunt grew weary of the big lights and, in 1881, withdrew from show business. Upon hearing rumours about 180 carat diamonds littering the Kalahari desert, he decided to swap the sawdust for desert sand and turn his attention to adventuring, with his male companion Lulu. It was during their journey through the desert that Hunt claimed to have made a startling discovery (or perhaps his greatest invention): the buried city of an ancient civilization.

The map opposite featured in the account of Hunt's expedition that he submitted to London's Royal Geographic Society (RGS) on his return. Note the small mark of 'Ruins' – Hunt writes that, while hunting, his party uncovered 'a huge walled inclosure [sic], elliptical in form and about the eighth of a mile in length'. This was, to his mind, the ruins of a city:

The masonry was of a cyclopean character; here and there the gigantic square blocks still stood on each other ... In the middle was a kind of pavement of long narrow square blocks fitted neatly together, forming a cross, in the centre of which was what seemed to be a base for either a pedestal or monument. We unearthed a broken column ... the four flat sides being fluted.

A sketch of the ruins allegedly discovered by Hunt.

Initially, the findings were met with little interest. The RGS members who commented on the Hunt report ignored the section on the ruins, and instead criticized the lack of information on the more pressing issue of local water sources.

Hunt then published a book of his adventure in 1886, under the title *Through the Kalahari Desert. A Narrative of a Journey with Gun, Camera, and Note-Book to Lake N'Gami and Back.* He went into further detail about the discovery, describing how his bemused local guides refused to help uncover the stones, thinking it pointless. The chapter, titled 'The Bastards Won't Dig', includes a diagram drawn by Lulu and a more detailed description of a 'long line of stone which looked like the Chinese Wall after an earthquake, and which, on examination, proved to be the ruins of quite an extensive structure.'

Hunt then ends with an ambiguous conclusion as to the identity of the relic, leaving it to 'others more learned on the subject' and summarizing the find with a poem:

A half-buried ruin – a huge wreck of stones
On a lone and desolate spot;
A temple – or a tomb for human bones
Left by man to decay and rot.

Rude sculptured blocks from the red sand project,
And shapeless uncouth stones appear,
Some great man's ashes designed to protect,
Buried many a thousand year.

A relic, may be, of a glorious past,
A city once grand and sublime,
Destroyed by earthquake, defaced by the blast.
Swept away by the hand of time.

A cynical mind might consider this a grubby attempt to manufacture a mystery for book sales and acclaim. If this were indeed the case, the author would have been disappointed, for there was little reaction to the book in the nineteenth century – perhaps because Hunt's reputation as a shrewd theatrical operator was so widespread. But then, in 1923, Professor E. H. L. Schwartz of Rhodes University resurrected the story of the buried civilization, and F. R. Paver, editor of the *Johannesburg Star*, and Dr W. Meent Borcherds were inspired to investigate the legend. Their articles caught the public

William Leonard Hunt, the Great Farini (1838–1929).

imagination, and expeditions were soon launched to find 'the Lost City of the Kalahari'. The number of attempts to find the ruins is difficult to quantify, but is certainly considerable – the historian A. J. Clement, writing in 1967, counts at least twenty-six missions by that time; and efforts continue: a 2010 international effort employed microlights to scour the savannah by air.

On the authenticity of the city, Paver concludes: 'The balance of probability is that Farini was either romancing – with or without some curious stones on which to build his fancies – or that he was describing a real building – a wildly improbable thing to find in those remote sand hills.' Indeed, it is suspicious that, of the many photographs taken by Lulu, none was of the ruins. Regardless, one can have little doubt that, somewhere, plans for another search for the Kalahari's lost city are being drawn up at this very moment.

MOUNTAINS OF KONG

FROM 11°36'N, 14°30'W TO 11°23'N, 4°49'E

In 1889, a man named Louis Gustave Binger stood
before an audience at the Paris headquarters of the
Société de Géographie and demolished an entire 3728-
mile (6000km) mountain range. The French explorer
had recently returned from a mission to the African
continent, tracking the path of the Niger River from Mali
to the kingdom of Kong. The latter, which is a real town
in the northern Ivory Coast, was known to be the home
of a chain of enormous mountains that led for countless
miles into the hazy distance. To Binger's great surprise,
far from finding himself in the shadow of such titans, he
discovered something quite different: 'On the horizon,
not even a ridge of hills!'

James Rennell's A Map
shewing the Progress of
Discovery & Improvement in
the Geography of North Africa
*(1798). It was the first map to
label the mountains with the
name 'Kong', though not the first
to show them, as is commonly
thought.*

For nearly a century, the Mountains of Kong were illustrated as running along the 10th parallel by cartographers attempting to fill the blank space of the continental interior while having very little data to go on. But to invent an entire mountain range, that would intimidate and prohibit future ventures of trade and exploration for years to come, would suggest that, at some point, a tremendous leap of theoretical mapping was taken. So who made it, and what prompted it?

James Rennell is most commonly accused of this cartographic crime, as his 1798 *A Map shewing the Progress of Discovery & Improvement, in the Geography of North Africa* is the first to feature the label 'Kong mountains'. The London mapmaker was an experienced and meticulous geographer – he had been lauded in 1779 for his *A Bengal Atlas*, a work instrumental in the service of British strategic interests – which makes the creation of a fiction as significant as the Kong Mountains all the stranger. Rennell's North Africa map

On his A New Map of Africa from the latest authorities *(1805), John Cary combined the Kong Mountains with the legendary Mountains of the Moon to create one giant belt stretching across the entire continent of Africa (see Mountains of the Moon entry on page 162).*

was one of two drawn to accompany Mungo Park's *Travels in the Interior Districts of Africa*, in which the fearless Scottish adventurer records his solo travels through Central Africa in search of the legendary city of 'Tambuctoo', equipped with little more than a compass, shotguns, a blue dress coat and a wide-brimmed hat. In the book, Park makes a key reference:

I gained the summit of a hill, from whence I had an extensive view of the country. Towards the south-east, appeared some very distant mountains, which I had formerly seen from an eminence near Marraboo, where the people informed me, that these mountains were situated in a large and powerful kingdom called Kong; the sovereign of which could raise a much greater army than the king of Bambarra.

Herisson's 1820 map of American mountain ranges, Carte de l'Amerique, *also shows the Kong Mountains in muscular form.*

Rennell considered this confirmation of his belief that the Niger began at this point in mountains and flowed to the west along the range. What is rarely mentioned, though, is that this theory was not of his invention. Since the sixteenth century, when Europeans first started penetrating the African mainland, there were rumours of giant mountain ranges in the region, which seemed to be confirmed by the Niger's path, and they can be found drawn, unnamed, on maps created earlier than Rennell's famous blunder – the *Mappe Monde* of Louis Denis, produced in Paris in 1764, is one such example.

Rennell would have been working in part from these maps, and would have found Park's mention as persuasive corroboration. In the appendix to Park's book, Rennell writes:

The discoveries of this gentleman … give a new face to the physical geography of Western Africa. They prove, by the courses of the great rivers, and from other notices, that a belt of mountains, which extends from west to east, occupies the parallels between 10 and 11 degrees of north latitude, and at least between the 2nd and 10th degrees of west longitude (from Greenwich). This belt, moreover, other authorities extend some degrees still farther to the west and south, in different branches …

After the publication of Rennell's map the mountains rose up on a series of other works – forty, in fact, have been found by the American academics Thomas Bassett and Philip Porter to bear the Kong falsity. Aaron Arrowsmith's *Africa Atlas* (London, 1802) brought them to the mainstream; Johann Reinecke drew the snow-covered slopes of 'Gebirge Kong' on his map of 1804; but most striking is John Cary's depiction on his 1805 *A New Map of Africa*, in which the Kong peaks are joined with the mythical Mountains of the Moon (see relevant entry on page 166), to form an impossibly vast transcontinental belt. In 1880, the German publication *Meyer's Conversation Guide* informed its readers that the Mountains of Kong were: 'unexplored mountains, extending north of the coast of Upper Guinea over a length of 800 to 1,000km between the seventh and ninth degrees latitude north, until longitude 1° west of Greenwich'. After featuring in Jules Verne's *Robur the Conqueror* in 1886, the Mountains of Kong then appeared on Rand McNally's 1890 map of Africa, and again on *Trampler's Mittelschulatlas* (Vienna, 1905). They made their final appearance in 1928, in the highly respected Bartholomew's *Oxford Advanced Atlas*, though even as late as 1995 they enjoyed a mention in *Goode's World Atlas*.

KOREA AS AN ISLAND

39°42'N, 126°24'E

Very little was known of Japan in late sixteenth-century Europe – murkier still was the land of Korea. The first map of Japan to be published in a European atlas, Ortelius's *Theatrum Orbis Terrarum* (1595), was a famously influential work that presents an early recognizable image of Japanese geography. It is based primarily on a map made in 1592 by Luís Teixeira, a Portuguese Jesuit cartographer, and was for many years the basis of cartography of the region until Martino Martini's more informed and elegant 1655 map was published by Joan Blaeu. At that time, it was the Portuguese who were the most informed of East Asia, having run a trade outpost on the island of Hirado since 1543.

Teixeira drew on Japanese sources for accurate information of that country's geography, but his portrayal of Korea is most striking to modern eyes, given that it is drawn as a carrot-shaped island named 'Corea insula' off the coast of China.

The first map of Japan in a European atlas, by Ortelius (1595).

In the north, Pyongyang is labelled 'Tauxem' and the southern point is marked 'Punta dos ladrones' (Cape of Thieves), below which are drawn the 'Isles of Thieves', reflecting rumours of a heavy pirate presence.

This insular notion was backed by Jan Huyghen van Linschoten's *Travel Accounts* of 1596, in which he observes: 'A little above Japan, on 34 and 35 degrees, not far from the coast of China, is another big island, called Insula de Core, from which until now, there is no certainty concerning size, people, nor what trade there is.' Hondius, too, produced a map of China in 1606, featuring Korea as an island. He also shows a sea monster and a depiction of a Japanese crucifixion, but it is the insular Korea that Hondius feels should be treated with caution, and so he added a note that there was some doubt as to whether it was, in fact, true. This inscription is included with later reproductions of his map, as well as others by seventeenth-century cartographers drawing an insular Korea, who seem to have ignored maps such as those by Diogo Homem in 1588 and João Teixeira (son of Luís) in 1630, which correctly identify Korea as a peninsula. Despite these contradictions, the idea of Corai Insula lingered: it can be found on a 1658 map by Johannes Jansson (which also shows the phantom 'Compagnies Land' published in his *Nieuwen Atlas*, and it crept onto maps until the eighteenth century (see Gamaland and Companies Land entry on page 120).

How Korea came to be thought of as an island by Europeans in the late sixteenth and seventeenth centuries is not known for sure, but the logical explanation is the same as that behind California's liberation from the mainland – incomplete circumnavigation finished with speculation. At the base of the peninsula's projection from continental Asia into the Pacific run two rivers that help serve as the border with China. To the west flows the Yalu River, which opens out into Korea Bay; and to the east runs the Tumen River, which pours out into the East Sea, or Sea of Japan. These rivers are broad and navigable – early explorers would have been able to travel some way along them and make the assumption that the two waterways joined together to form a continuous strait around the north of the peninsula, thus rendering it an island. Had they travelled further, though, they would have encountered Mount Paektu, Korea's highest peak and the source of both rivers, and in the shadow of this giant mountain range the mistake would no doubt have been hastily rubbed out.

Hondius's annotated depiction of Korea as an island (c.1619).

LOST CONTINENTS
OF LEMURIA AND MU

Philip Sclater, secretary of the Zoological Society of London for forty-two years, was a foremost expert in the field of ornithology in the nineteenth century. Oxford-educated and a fellow of the Royal Society, he wrote more than one thousand papers, books and articles. In his 1858 paper published by the *Proceedings of the Linnean Society* Sclater divided the world into six zoological regions, which remain in use today: Aethiopian, Australasian, Indian, Nearctic, Neotropical and Palaearctic. Seven animals have been named in his honour, including the Mexican chickadee (*Poecile sclateri*), the erect-crested penguin (*Eudyptes sclateri*) and Madagascar's blue-eyed black

Below and opposite: Maps of the theoretical continent of Lemuria, from W. Scott-Elliot's The Story of Atlantis and Lost Lemuria *(1925) published by the Theosophical Publishing House.*

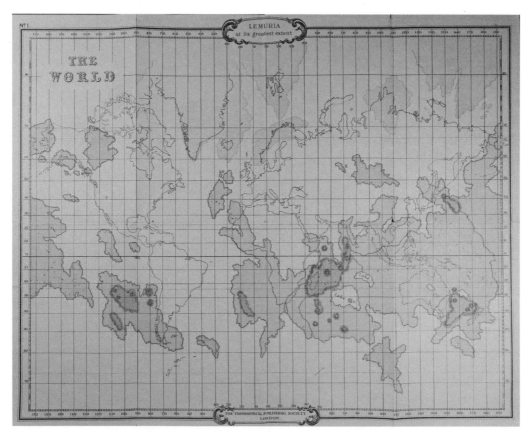

lemur (*Eulemur flavifrons*), which is known as Sclater's lemur. And yet it is for his championing of one particularly strange theory – the existence of a sunken continent named 'Lemuria' – that Sclater is best remembered.

In 1864, Sclater submitted an article to *The Quarterly Journal of Science,* positing a theory as to the puzzling similarities in primate fossils found in both Madagascar and India, despite their being divided by the Indian Ocean:

The anomalies of the Mammal fauna of Madagascar can best be explained by supposing that … a large continent occupied parts of the Atlantic and Indian Oceans … that this continent was broken up into islands, of which some have become amalgamated with … Africa, some … with what is now Asia; and that in Madagascar and the Mascarene Islands we have existing relics of this great continent, for which … I should propose the name Lemuria!

This idea of a 'stepping-stone continent' came well before the acceptance of plate tectonics and continental drift (Alfred Wegener's landmark book *The Origin of Continents and Oceans*

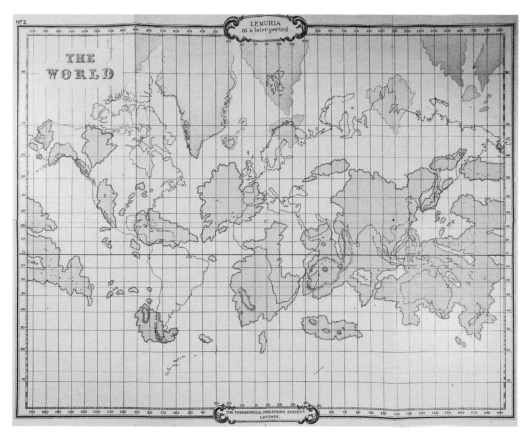

would not be published until 1915), and this lost 'Lemuria' continent was supported by some in the scientific community, such as the Darwinian taxonomist Ernst Haeckel, who suggested Lemuria could also explain the absence of the so-called 'missing link'.

Though it was superseded by advances towards the tectonic idea, Lemuria quickly gained a second life thanks to its popularity with occultists. The eccentric Madame Helena Blavatsky, the Russo-American co-founder of the Theosophical Society, took the idea further by writing, in 1885, that not only had the continent existed but that it had also taken up almost the entire southern hemisphere 'from the foot of the Himalayas to within a few degrees of the Antarctic Circle', and was inhabited by a race of lost people, the Lemurians. These were one of the seven 'root-races of humanity' – half-human, egg-laying hermaphrodites with an average height of 7ft (2.1m), who bred with animals. They were also non-corporeal, existing on the astral plane. Blavatsky was made privy to this information, she said, by the Mahatmas, who revealed their ancient Indian wisdom-writings in the pre-Atlantean *Book of Dyzan*. According to Blatavsky, the Lemurian continent was destroyed by the gods in the Third Eocene Age and its people swept away, their descendants being the doomed Atlanteans – and also the Australian Aborigines, Papuans and Hottentots.

These ideas were later published by the English author William Scott-Elliot in *The Lost Lemuria* (1904), after receiving added material from fellow Theosophical Society member Charles Webster Leadbeater, who was helpfully able to communicate with the Theosophical Masters via 'astral clairvoyance'. This book, later published together with the story of Atlantis (see relevant entry on page 24), was accompanied by maps depicting the full extent of the Lemurian continent overlaid on modern maps, shown on pages 152–3. As to the source of information for these maps, Scott-Elliot would say only that the Atlantean maps were made by 'mighty Adepts in the days of Atlantis', and the Lemurian charts 'by some of the divine instructors in the days when Lemuria still existed'.

Another theoretical lost continent named 'Mu' also emerged in the nineteenth century and is often confused with Lemuria. A French physician named Augustus le Plongeon, in his book *Queen Móo & The Egyptian Sphinx* (1896), claimed to have translated ancient Mayan writings using his knowledge of Egyptian hieroglyphics, and the 'De Landa alphabet',

Helena Blavatsky in 1889.

a translation of Mayan glyphs with corresponding Spanish letters created by the sixteenth-century bishop of Yucatán, Fray Diego de Landa, as part of his documentation of the Maya civilization. Le Plongeon writes of discovering references to 'the existence, destruction, and submergence of a large island in the Atlantic Ocean' – the 'Land of Mu', he suggests, was the true name of Atlantis. Le Plongeon's book was later followed by Colonel James Churchward's *The Lost Continent of Mu, the Motherland of Man* in 1926, in which the American writer theorizes that Mu had originally stretched across the Pacific, similar to the geography of Lemuria, until it 'vanished in a vortex of fire and water 12,000 years ago'. This, he said, he had learnt in India fifty years previously from a high-priest.

Today, the continents of Lemuria and Mu are dismissed as impossible – zany relics of outlandish theory that say more of their authors than their geography. Researchers who have attempted to use the de Landa alphabet employed by le Plongeon have been forced to conclude that the system was a spurious invention of the bishop, as all it has ever produced is gibberish.

MARIA THERESA REEF

36°50's, 136°39'w

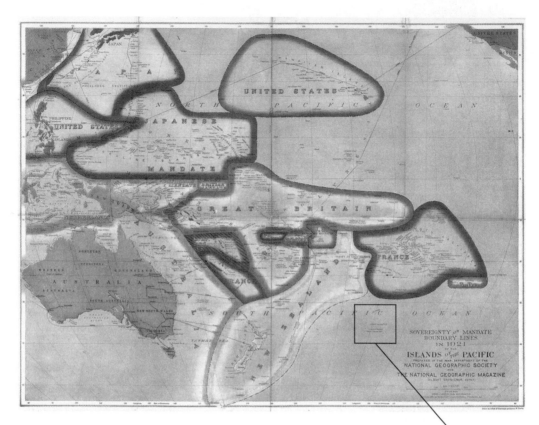

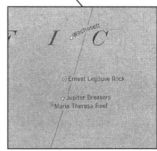

In Jules Verne's *The Children of Captain Grant* (1867), the titular character finds shelter on a solitary island exactly on the 37th parallel: 'a little isle marked by the name of Maria Theresa, a sunken rock in the middle of the Pacific Ocean, 3,500 miles from the American coast, and 1,500 miles from New Zealand'. In 1874, Verne writes about the place again in *The Mysterious Island*, referring to it as 'Tabor Island', which is how it is marked on French charts. It is a perfectly reasonable setting for the author to choose, as the reef decorated charts available to Verne at the time, having been discovered only thirty years earlier on 16 November 1843. The *Maria Theresa,* a whaler originating from New Bedford, Massachusetts, glimpsed a reef breaking the waves, and Captain Asaph P. Taber recorded its location. No other skipper ever laid eyes on

Sovereignty and Mandate Boundary Lines of the Islands of the Pacific in 1921, *made for the National Geographic Society, showing the nonexistent Maria Theresa Reef. Its neighbours – Ernest Legouvé Rock, Jupiter Breakers and Wachusett Reef – are also phantoms.*

it, but over the years Maria Theresa Reef eluded purging. Vessels passed over its coordinates without difficulty, but no one wanted the responsibility of erasing a possible danger to shipping that might well exist at a slightly different position.

This half-existence of Maria Theresa Reef went on for decades. In 1966, *CQ Amateur Radio* magazine published an account by radio enthusiast Don Miller, who staged a live broadcast from the shore of Maria Theresa Reef. It showed a photograph of Miller setting up his radio equipment on the reef, the sea lapping at his shins. He wrote that he had been forced to remain on his chair the entire time, for fear it would be swept away. By this time, Maria Theresa Reef had the reputation of a phantom after a 1957 search found nothing, and so Miller was ridiculed. 'I was there, John', he protested to his friend John Steventon, 'and I operated there. As we approached the reef it came up out of the ocean and we landed and operated. And after we had finished and were leaving, we looked back and saw the island sinking back into the sea. And that positively was the way it was. Absolutely!'*

Yet when the New Zealand oceanographic research ship HMNZS *Tui* later examined the area of the supposed Maria Theresa Reef, the crew found the waters to have a depth of 2734 fathoms (5000m). But Hydrographic Office chart No. 2683, issued in 1978, has Maria Theresa listed; and in 1983 the Hydrographic Office recalculated its position again, from 151.13 degrees west to 136.39 degrees west, more than 620 miles (1150km) further east from Captain Taber's original claimed location. On maps, Maria Theresa is usually depicted amid a group of other large reefs in that area, namely Jupiter Reef, Wachusett Reef and Ernest Legouvé Reef. All of them are phantoms, though they occasionally crop up in modern publishing: Ernest Legouvé, for example, can be found in the 2005 edition of the *National Geographic Atlas of the World*, and – at the time of writing – is placed by Google Maps at 35°12'S, 150°40'W.

---

* *Incidentally, Don Miller's Maria Theresa 'QSL' card, which radio enthusiasts use to confirm a successful transmission from their location, has since become a prized curio for collectors.*

MAYDA

52°00'N 33°45'W   *Also known as Maida, Mayd, Mayde, Brazir, Mam, I. Onzele,*
Asmaida, Asmaidas, Asmayda, Bentusla, Bolunda and Vlaenderen, I. man orbolunda

Tracing the cartographic history of the island of Mayda is like tracking a spy through a series of forged identities, although, as it moves around the North Atlantic over the years, adopting a range of names and shifting in shape, it never quite escapes recognition. Mayda is one of the oldest and most enduring of phantoms, stubbornly clinging to the skins of maps for more than five centuries; it was the last of the mythical North Atlantic islands to be expunged. But in a strange twist, it may be that the phantom label is too readily applied.

Blaeus's map of the Americas from 1649 shows 'As Maydas' in the top-right corner'.

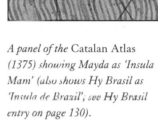

A panel of the Catalan Atlas *(1375) showing Mayda as 'Insula Mam' (also shows Hy Brasil as 'Insula de Brazil'; see Hy Brasil entry on page 130).*

In its signature crescent shape, Mayda first appears on the Pizzigani map of 1367 with the label 'Brazir'. One would, of course, assume this to refer to Hy Brasil (see relevant entry on page 130), if it weren't for the fact that this island is also marked in an early form to the west of Ireland with the warning 'Harmful'. This seems to suggest that the island to be known as Mayda was born out of confusion, a duplication that took on a life of its own. On the Pizzigani map, 'Brazir' carries its own warnings to sailors, with three Breton ships depicted under attack by sea monsters, one being dragged under by a giant octopus while a dragon flies overhead with a man in its jaws. When next seen on the Catalan map of 1375, Brazir is now 'Mam', for reasons one can only speculate – some have suggested that it might be based on a sighting by Irish sailors naming it after the Isle of Man, but there is nothing to support this idea.

No sooner did the ink dry on this incarnation than the island continued its transformation. On the Pinelli map of 1384 it is drawn with the name 'I. Onzele'; and, in 1448, Andrea

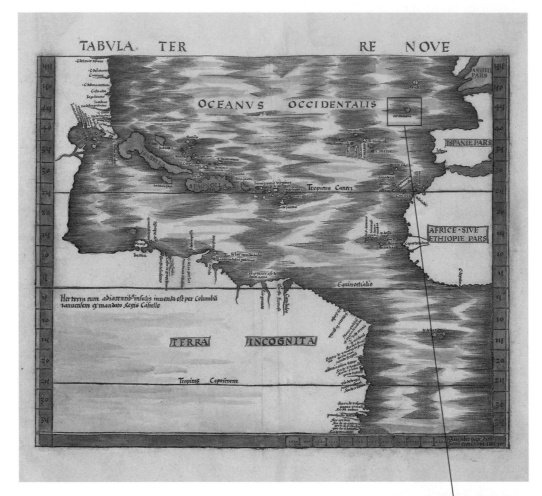

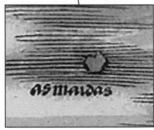

Bianco labels it 'Bentusla'. Other maps around this time also feature it but leave it nameless, partnering it with Hy Brasil, the two drifting around the North Atlantic in tandem. In 1513, Martin Waldseemüller includes it on his depiction of Ptolemy's *Geography*, ditching the crescent shape for a rough circle with a northwest chunk removed. The famously assiduous German cartographer names it 'Asmaidas', the first recognizable root of its later derivative. 'Asmaidas' is then trimmed by an anonymous Portuguese mapmaker in 1520 to 'Mayd', and the Prunes map of 1553 then sets the standard with 'Mayda'.

Though general consensus on the name had been reached, its location varied wildly; some even began to tow Mayda across the Atlantic to the North American coast. The Nicolay map of 1560 places it parallel to northern Newfoundland, and gives it the bizarre name 'I. man orbolunda', which might have some connection to its original crescent-moon shape. Mercator and Ortelius, the sixteenth-century's most respected

Martin Waldseemüller's groundbreaking map of 1513, Tabula Terre Nove, *is one of the earliest maps to focus on the New World. Commonly known as the* Admiral's Map, *it was included in Johann Schott's edition of Ptolemy's* Geographia, *published in Strasbourg in 1513. Mayda can be found as the green-coloured 'Asmaidas' in the top-right corner.*

cartographers, ignore Mayda, which goes some way to suggest there was little in the way of contemporary evidence for its existence, though they do draw a curved island in a similar position west of Brittany and name it 'Vlaenderen', which is recognizable as Flanders. This name was ignored by later cartographers, who continued to employ 'Mayda' for the next three hundred years, including it on nearly every map that featured the Atlantic Ocean.

A life of centuries, and yet we have no recorded accounts of anyone having set foot on the island's shores. So why was the case of Mayda such a convincing one to the greatest and most diligent of mapmakers? The answer could be that the island wasn't as spectral as we have assumed, for in 1948 something remarkable happened: a ship named the *American Scientist*, on its way from New Orleans to London, found itself at 46°23'N, 37°20'W, south of Greenland and – importantly – due west of southern Brittany. For some unknown reason, the captain of the freighter decided to measure the depth, perhaps noticing a variation in water colour. According to the charts, the water at this position was supposed to be 2400 fathoms (4390m), but the *American Scientist*'s sonar gave readings of only 20 fathoms (36.5m). The crew performed another test and confirmed that, below their ship, in the middle of the Atlantic, there lay a raised point of land 28 miles (45km) in diameter. A second ship, the *Southland*, confirmed the findings and even reported the submerged land to have a bay indented in its northern end. Through some act of geologic violence centuries ago it seems possible that the island of Mayda entirely vanished beneath the waves – a few brushstrokes of ink on old maps being the only remaining indication that an entire island once stood proud of the Atlantic at that spot.

Mayda is here marked as 'Isola de Maydi' on the Prunes map of 1553.

MOUNTAINS OF THE MOON

00°23'N, 29°52'E   *Also known as Mone Lune, Montes Lunae, Al Komri*

Even when virtually nothing was known of Africa by Europeans, whose maps of the continent were little more than a ghostly outline of blank space brushed with a few token strokes of river, one giant feature was depicted from the beginning: the Mountains of the Moon, a mythical range said to be the source of the Nile.

The origin of that river had been speculated for thousands of years. In the fifth century BC, Herodotus travelled to Egypt and interviewed locals for details of their culture and geography. He writes in his grand 'enquiry', *The Histories*, that, among all the Egyptians, Libyans and Greeks he met, he spoke to only one who claimed to know the source of the Nile, a scribe who kept the register of the sacred treasures of Minerva in the city

The revised second state of Mercator's Universalis Tabula Iuxta Ptolemaeum *shows the* Lunae Montes.

of Sais and whose story struck him as truthful (though he remained sceptical). In Book 2 Chapter 28, Herodotus quotes the man as saying: 'Between Syene, a city of the Thebais, and Elephantine, there are two hills with sharp conical tops; the name of the one is Crophi, of the other, Mophi. Midway between them are the fountains of the Nile, fountains which it is impossible to fathom. Half the water runs northward into Egypt, half to the south towards Ethiopia.' These fountains had been proven to be bottomless by the Egyptian king Psammetichus, who ordered a rope to be made many thousand fathoms in length and lowered into the fountain, but the floor was never reached.

Five hundred years later, in his description of Aethiopia in Book IV, 8 of *Geographia*, Ptolemy mentions the 'Barbarian Bay', where the 'Aethiopian Anthropophagi dwell', a vicious cannibalistic race, 'and from these towards the west are the Mountains of the Moon, from which the lakes of the Nile receive snow water'. This is corroborated by the account of Diogenes, who travelled inland for twenty five days from Rhapta (a port now lost) in east Africa and describes witnessing the Nile emerge from giant mountains, which he reports the natives refer to as 'the mountains of the moon' due to the gleaming whiteness of their snow-covered peaks.

Alexander the Great and Julius Caesar both considered launching expeditions in search of the wonders promised by these stories. Later Arab geographers, such as Abu'l-Fida, and Muhammad al-Idrisi of the twelfth century, also took these accounts as reliable sources, the latter writing: 'These two parts of the Nile spring from the mountains of the *moon*, which are situated 16° beyond the equator. From these mountains, the Nile issues in 10 streams, five of which flow together into one great lake, and the remainder into another such lake.'

On Waldseemüller's 1513 map of southern Africa, the first separate map of the area to be produced, the 'Mone Lune' are there (together with, in the lower right corner, the king of Portugal grasping a sceptre and the royal banner of Portugal while riding a sea monster, representing the expansion of Portugal's mercantile empire under his reign). On another early European map of Africa, Sebastian Münster's *c.*1550 'Totius Africæ tabula …', this system of rivers pouring from the *Montes Lunae* into two large lakes is also shown, the bodies draining into the Nile. (Other mythical embellishments to look for on this map include a monstrous monoculi seated over

Nigeria and Cameroon; the kingdom of Prester John; and 'Mero', the mythical tombs of the Nubian kings; see Kingdom of Prester John entry on page 194.)

In the seventeenth century, the mountains and their depthless fountain were drawn by Athanasius Kircher, whose map of Atlantis is shown on page 25. The German scholar created this subterranean depiction from the description of his contemporary, Pedro Páez, a fellow Jesuit who visited Ethiopia at the beginning of the century and recorded seeing the 'fountains of the Nile'. His original report was translated and reproduced by James Bruce, the Scottish explorer of the Nile who spent more than twelve years in Africa, in *Travels to Discover the Source of the Nile* (1790):

*On the 21st of April, in the year 1618, being here, together with the king and his army, I ascended the place, and observed every thing with great attention. I discovered first two round fountains, each about four palms in diameter, and saw, with the greatest delight, what neither Cyrus king of the Persians, nor Cambyses, nor Alexander the Great, nor the famous Julius Caesar, could ever discover. The two openings of these fountains have no issue in the plain on the top of the mountain, but flow from the root of it. The second fountain lies about a stone-cast west from the first: the inhabitants say that this whole mountain is full of water, and add, that the whole plain about the fountain is floating and unsteady, a certain mark that there is water concealed under it; for which reason, the water does not overflow at the fountain, but forces itself with great violence out at the foot of the mountain …
[T]he fountain seems to be a cannon-shot distant from Geesh …*

At the opening of the nineteenth century, the myth of the Mountains of the Moon being the source of the Nile was much in doubt, though this didn't prevent the normally assiduous John Cary from producing his bizarre 1805 map (see page 151), in which he links the Mountains of Kong with the Mountains of the Moon to form one giant range across the entire continent. The British explorers John Hanning Speke and Richard Francis Burton set off in 1856 to find the source of the Nile, a constrasting pair who quarrelled from the outset. It was a gruelling, three-year mission into the interior of Africa, in which the pair suffered all manner of tropical diseases: several times Burton fell gravely ill and Speke went temporarily blind – he also lost his hearing for a time when a beetle crawled into his ear and he tried to fish it out with a knife. When Burton became too ill to move, Speke continued solo and identified Lake Victoria as the true source in 1858 (much to the disbelief and protest of Burton). The pair feuded for years until in 1874 Henry Morton Stanley (deliverer of the famous and probably invented line 'Doctor Livingstone I presume?') circumnavigated the lake and confirmed Speke's findings.

This finally laid the Moon Mountains myth to rest, but the discussion was now as to which mountains served as its basis. In 1940, the writer G. W. B. Huntingford made the case for the range to be identified with Mount Kilimanjaro, but was ridiculed by his peers, though Sir Harry Johnston had made the same argument in 1911, as did Dr Gervase Mathew in 1963. Today, it is thought that, if the descriptions of the mountains weren't totally fabricated, the snow-capped Rwenzori Mountain range in the Democratic Republic of Congo are the most likely basis for the myth, due to their corresponding location in eastern equatorial Africa.

Martin Waldseemüller's Tabula Nova Partis Africae *(1541).*

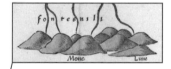

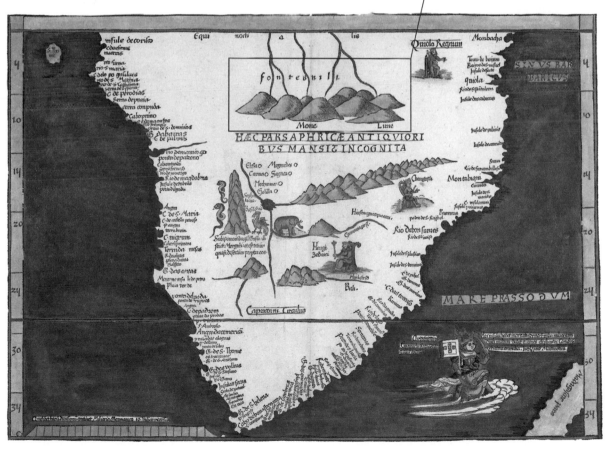

LANDS OF
BENJAMIN MORRELL

The exuberant exploration accounts of Benjamin Morrell (known commonly as 'the biggest liar of the Pacific') were plagiarized so liberally from the travel writing he obsessively collected that it is hard to discern, in the words of oceanographer Henry Stommel, 'where the quotation ends and Morrell's own experience begins'.

Born in 1795 in Westchester County, New York, by the age of seventeen he had run away, with dreams of adventure at sea. He sailed for years 'before the mast' (as a lowly seaman, rather than as an officer). During the War of 1812, which broke out over numerous tensions between the United States and Great Britain, he was captured twice by the British: his first detainment lasted for eight months, while the second saw him thrown into England's Dartmoor prison for two years. Upon release he returned to a life at sea, but his fortunes improved when the Quaker Captain Josiah Macy took him under his wing and taught him the skills of navigation. In 1821, he secured an appointment as chief mate aboard the *Wasp*, captained by Robert Johnson, and was thrilled to find they were to embark on a journey to the storied South Shetland Islands, discovered only three years earlier by the the British voyager William Smith. After an adventurous year surviving gales, near-drownings and his ship becoming trapped in ice, Morrell returned to New York where he was awarded captaincy of the *Wasp*. He set his sights on leading famous voyages of his own, and ultimately to produce a book that might bring him a level of acclaim similar to that of his heroes. *A Narrative of Four Voyages …* by Benjamin Morrell was published in 1832, written apparently after the fact, with unrestrained exaggeration. Works of this kind are commonly littered with self-aggrandisement and sexed-up narrations, but Morrell's stands out, thanks to his knack for discovering and exploring islands that simply didn't exist.

Benjamin Morrell, 'the biggest liar in the Pacific'.

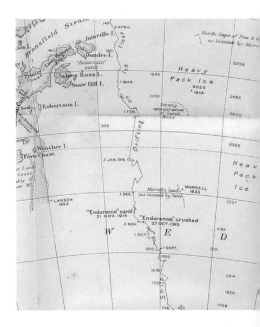

Map of The Voyage of the Endurance *from Shackleton's* South *– the account of his 1914–17 Antarctic journey. 'Morrell's Land (as located by him)' is marked in the centre.*

NEW SOUTH GREENLAND

67°52's, 44°11'w   *Also known as Morrell's Land*

The first of the illusory Morrell lands was New South Greenland, which was sighted during the first of the four voyages for which he is known. In 1822, commanding the 123-ton schooner *Wasp*, Morrell embarked on a sealing expedition to the Southern Ocean by way of Rio de Janeiro and the coast of Patagonia. While heading south, he spotted Bouvet Island (with 'improbable ease', in the words of historian William Mills), and made a doubtful claim to being the first to land on its shores. He named it in honour of Jean-Baptiste Charles Bouvet de Lozier, who sighted it on 1 January 1739 and recorded its coordinates with such inaccurate dead reckoning that none, including Captain Cook, was able to find it until the British whaler captain James Lindsay spotted it in 1808 and named it Lindsay Island. In his journal, Morrell fails to make any reference to Bouvet's distinctive feature – it being covered almost entirely by a glacier that fills the crater of an inactive volcano – and so his version of events is held with deep suspicion. It is also strikingly similar to a description given by George Norris, who landed there in 1825.*

After a hunting expedition that yielded 196 seals, Morrell moved on to the Kerguelen Islands and the South Sandwich Islands, and then wrote of penetrating the Weddell Sea, off the northwestern coast of the Antarctic. This was a region entirely unexplored. Here, on 19 March 1824, Morrell sighted 'the North Cape of New South Greenland', after three days of no observations. The land 'abounds with oceanic birds of every description; we also saw about three thousand sea-elephants, and one hundred and fifty sea-dogs and leopards'.

---

* *Incidentally, Bouvet is the setting for a most peculiar mystery. In 1964, a South African supply ship and Royal Navy ice vessel HMS* Protector *were dispatched to the island to judge its suitability for housing a weather station. It was there, in one of the remotest places on Earth, that Lieutenant Commander Allan Crawford spotted an unmarked, abandoned lifeboat among a colony of seals in a lagoon. He wrote: 'What drama, we wondered, was attached to this strange discovery. There were no markings to identify its origin or nationality. On the rocks a hundred yards away was a forty-four gallon drum and a pair of oars, with pieces of wood and a copper flotation or buoyancy tank opened out flat for some purpose. Thinking castaways might have landed, we made a brief search but found no human remains.' Bouvet was visited again in 1966 by a biological survey team, but no mention of the vessel was made, and it was never found again. The lifeboat at the end of the world remains an unsolved mystery.*

Given Morrell's reputation for deception, the inclination is to dismiss New South Greenland as deliberate invention. Indeed, the historian Raymond Howgego points out that some of his description here is plagiarized from James Weddell's *A Voyage towards the South Pole*. But there is little self-aggrandising in this part of Morrell's account, and it is entirely possible that it was born out of error, either through Antarctic mirage (as was later claimed by Wilhelm Filchner of the survey ship *Deutschland*, in 1912) or perhaps confusion with distant icebergs, combined with some faulty positioning.

BYER'S ISLAND

28°32'N, 177°4'E

On his return to New York, Morrell married Abby Jane Wood in June 1824 and attempted to settle into domesticity, but just one month later he was off again on another two-year voyage in the *Tartar*. In April 1825, he reached California, by way of Buenos Aires and the Galapagos, spending only a short time there before heading for the Hawaiian Islands. He sailed west along the island chain, crossing the 180th meridian, and on 12 July 1825 recorded an encounter with an island that he named 'Byer's Island', in order to flatter the New York shipowner and merchant James Byers:

We landed on Byer's Island, situated at lat. 28°32' north, long. 177°4' east [he reported]. This island is moderately elevated, and has some bushes and spots of vegetation. It is about four miles in circumference, and has good anchorage on the west-south west side, with fifteen fathoms of water, sand and coral bottom. There are no dangers around this island, excepting on the south east side, where there is a coral reef, running to the southward about two miles. Sea-birds, green turtles, and sea-elephants resort to this island; and a plenty of fine fish may be caught with hook and line about its shores. Fresh water may be had here from the south-south west side of the island, which is of volcanic origin.'

This, too, does not exist. On the same day, Morrell also recorded another land that no one has since been able to find, which this time bore his own name.

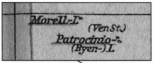

A map produced in Leipzig c.1890 shows 'Morrell I.' as well as 'Byers I'.

MORRELL'S ISLAND

29°57'N 174°31'E

We then hauled on a wind to the north [Morrell wrote], the water being perfectly smooth; and after running along under the lee of the reef at the rate of seven miles an hour, for two hours, on a north-by-west course, we saw the land from the mast-head, bearing north west. We immediately kept off for it, and at 10 a.m., we were close in with a small low island, covered with sea-fowl, and the shores of which were lined with sea-elephants. Green turtles were found here in great abundance, and two hawk's-bill turtles were seen. This island presents all the usual indications of volcanic origin.

On the west side of the island, Morrell reported a coral reef of about 15 miles (28km), another on the southeast side that extended about 30 miles (56km), that were useful for anchorage. The island was low, he said, being nearly level with the surface of the sea, and about 4 miles (7km) in circumference. After examination, he found the island afforded neither furs nor other valuable articles, and so he 'left it to its solitude, and steered to the north on a wind'.

Despite these discoveries, which were most likely engineered to add some distinction and excitement, the New York sponsors behind Morrell's second voyage were unhappy with the results of the voyage, and it was two years until he was able to secure backing for a third command.

Along with Byers's Island, Morrell's Island remained on navigational charts for more than one hundred years, surviving the great cull of 123 islands (including three real ones) from official charts by the British Admiralty in 1875. Between 1907 and 1910, the International Date Line was actually redrawn around Morrell's Islands to place them in the same time zone as Hawaii.

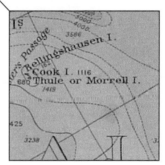

E. P. Bayliss and J. S. Cumpston's Handbook and Index to accompany a Map of Antarctica *(1939).*

NORUMBEGA

44°45'N, 70°17'W

Also known as Norombega, Aranbega, Oranbega

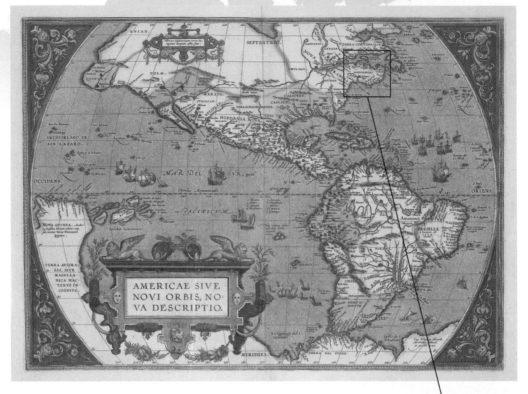

The much-debated origin of Norumbega dates back to
the voyage of Giovanni de Verrazano in the sixteenth
century. After he returned from his famous journey
along the North American coast in 1524, the explorer
submitted a report of his findings to the French king,
Francis I. Further details he disclosed to a cartographer,
Visconte Maggiollo, who published a map in 1526
exhibiting these details – among which was the
Verrazano discovery of a 'Norman villa' in the area
of modern New England. Meanwhile, Verrazano's
brother Giralamo, a mapmaker and another confidant,
was working on his own map, which was printed in
1529 and in which he represented a river with the label
'Norombega', the first recognizable instance, in the
region now known as Maine. The two were swiftly
linked by their similar form, and, though it was initially

*Norumbega on Ortelius's 1570
map of America. (Also shows
Friesland, Estotiland, Isle of
Demons and St Brendan's Isle;
see Phantom Lands of the Zeno
Map entry on page 240, Isle of
Demons entry on page 84 and
St Brendan's Island entry on
page 202.)*

applied to a river, 'Norumbega' came to refer to an area of land, which was first inked as such in 1542 on the Euphrosinius Ulpius globe. At this time, Jacques Cartier was making his final voyage along the North American coast, during which his navigator, Jean Alfonce, recorded the 'Cape of Norumbegue' (clearly Cape Cod) and, to the west, the 'River of Norumbegue', which seems to be the Narragansett. Some 40 miles (64km) north, writes Alfonce, was the city of Norumbega, 'and there is in it a good people, and they have pelts of all kinds of animals'.

For the next sixty years, with little further French exploration of the area, Norumbega was included on maps on the east coast of 'New France', as shown on the map on page 174 by Ortelius, which features the spires of the great fortified city.

To add to this cocktail of truth and confusion, we also have the wholly unreliable account of David Ingram, a sixteenth-century English sailor and explorer who claimed to have walked 3000 miles (4830km) across the interior of the North American continent in eleven months, from Mexico to Nova Scotia, in 1568. Ingram references the city in his account of

Map of the Americas *by Willem Blaeu (1617)*.

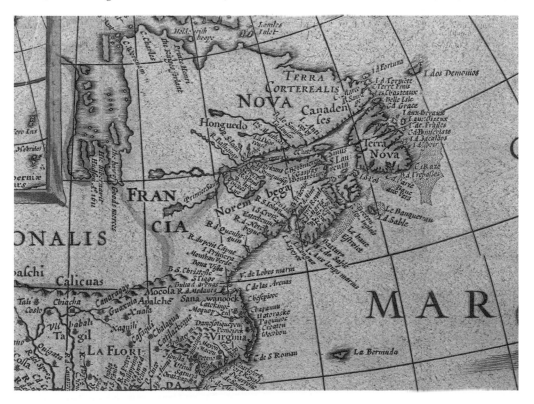

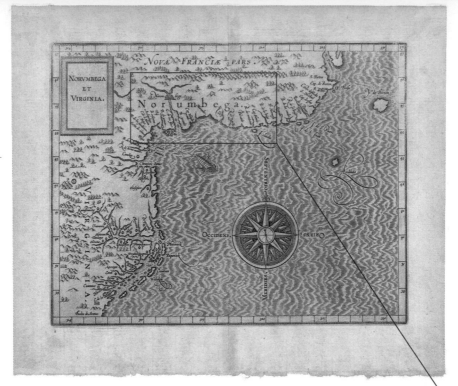

the journey, which was written thirteen years later, in 1582, by Sir Francis Walsingham – for Ingram himself was illiterate – and was released in 1589 in Richard Hakluyt's *The Principall Navigations, Voiges and Discoveries of the English Nation*. (In the second edition, Hakluyt omits Ingram's story – Samuel Purchas noting of this that: 'It seemeth some incredibilities of his reports caused him to leave him out in the next impression, the reward of lying [being] not to be believed in truths.')

Corneille Wytfliet Norumbege et Virginie *(1597)*.

Of the peoples he encountered, Ingram writes: 'Genirallye all men weare about there armes dyvers hoopes of gold and silver which are of good thickness,' and 'The women of the country gooe aparyled with plates of gold over there body much lyke unto an armor.' He then came across the city in question: 'The towne [is] half a myle longe and hath many streates farre broader than any streate in London,' he says. 'There is a great aboundance of gold, sylver and pearle, and … dyvers peaces of gold some as bigge as [my] finger, otheres as bigge as [my] fist.'

---

* *Gilbert was forced to abandon settlement plans due to a lack of supplies, and perished on the return journey, after stubbornly refusing to transfer from his small ship, the* Squirrel, *to a hardier vessel. Encountering waves 'breaking short and high Pyramid wise', according to Edward Hayes, a captain in his fleet, Gilbert was observed on deck crying: 'We are as near to Heaven by sea as by land!' before he, his ship and all souls aboard went down.*

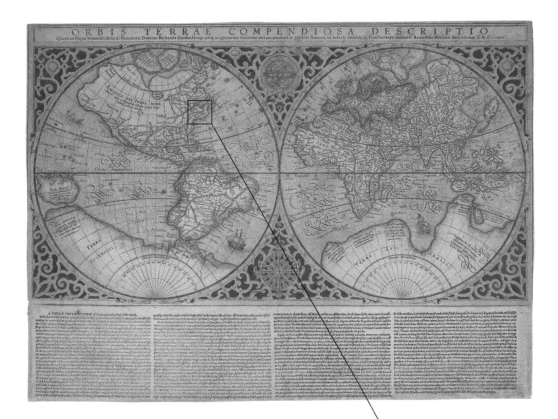

By this time, Norumbega had become keenly sought after. Ingram returned to the New World in 1583 with Sir Humphrey Gilbert in his unsuccessful and ill-fated* attempt to seize 9 million acres (3.6 million hectares) around the river Norumbega and establish an English settlement in Newfoundland. The French were also eager to find out more. In 1603, King Henry IV sent Samuel de Champlain to reconnoitre this perceived Eden. Though his searching for Norumbega resulted in the colony of Port Royal (Annapolis Royal, Nova Scotia), de Champlain was left disappointed. In the scornful words of Marc Lescarbot, his recordist: 'If this beautiful Towne hath ever been in nature, I would know who hath pulled it downe: For there is but Cabins here and there made with pearkes [poles], and covered with barkes of trees, or with skinnes …'

From this point, the French labelled a native settlement at the mouth of the Penoscot as 'Norumbega', and interest began to fade, evidenced by its diminished appearance on maps. Strangely though, the Dutch cartographers remained fascinated, and Norumbega can be found on works by the Holland mapmakers into the eighteenth century.

A 1616 edition of Mercator's landmark Orbis terrae compendiosa description.

CREATURES OF THE
NUREMBERG CHRONICLE MAP

The *Liber Chronicarum*, known as the *Nuremberg Chronicle*, of Hartmann Schedel, printed in Nuremberg by Anton Koberger in 1493, is a complete history of the Christian world from the time of the Creation up to the age of the book's production. Divided into eleven world ages, the history is informed by a variety of classical and medieval sources, from Pomponius Mela and Pliny the Elder to the Venerable Bede and Vincent of Beauvais. One of the most profusely illustrated of all German incunables, there is a map nestled among its pages that stands out from its contempories because of its stunning depictions of – to borrow a phrase from Wordsworth – a parliament of monsters, believed to inhabit distant lands of the known world.

The reverse side of the Liber Chronicarum *by Hartmann Schedel (1493). See pages 176–7 for the map itself.*

The first edition of the book was printed in Latin; a later one, in German, was slightly simplified and geared towards the less-educated reader. As with other medieval maps, one of the heavily drawn-upon sources, especially for the illustrations, is the *Collecteana rerum memorabilium (Collection of Remarkable Facts)* by Gaius Julius Solinus (*fl.* AD 250), who was known as 'Pliny's Ape' by his detractors for shamelessly borrowing nonsensical elements from the great writer's *Natural History*, an encyclopedia of ancient knowledge. Solinus's vivid portrayals of the monstrous races described by Pliny, Mela and others were an instant success and remained popular for a millennium. Though 'guilty of many untruths' (as Albertus Magnus describes him), Solinus was hugely influential on later depictions of the range of bizarre hominoid species rumoured to exist in the shadowed expanse beyond the known world.

The *Nuremberg Chronicle Map*, here from the original Latin edition, is perhaps the finest example of this mythological portraiture. In three corners of the design are the sons of Noah, who, according to the Scriptures, divided the world after the Flood: Shem and his descendants took Asia; Japheth, Europe; and Ham, Africa. Another item of interest is the mythical island of Taprobana, painted to the east across the Indian Sea (see Tapobrana entry on page 220). But it is the gallery of creatures sat to the far left of the map, together with a double panel on the reverse side, that are the most fascinating. Each possesses uniquely abnormal biology, but is unlabelled, so which creatures of legend do the illustrations represent?

A This character is commonly referred to only as the 'six-handed man'. In the histories of Alexander the Great (stories which were later deemed as a masquerade written by an author referred to as Pseudo-Callisthenes), this poly-limbed group of people is said to live in India.

B – The Gorgades These women covered in hair are presumably the same creatures written about by Pliny, which he labels Gorgades ('Gorgon-like ones'). Supposedly, they lived on islands scattered throughout the Atlantic. Pierre d'Ailly in *Imago Mundi* describes them: 'The Gorgodes Islands of the ocean … are inhabited by the Gorgodes, women of destructiveness, with coarse and hairy bodies.' Homer, too, mentions hairy 'Gorillae' women. Columbus took both the works of Pliny and d'Ailly with him on his first voyage, but seems to have been much keener on finding the Amazons over the hirsute female race.

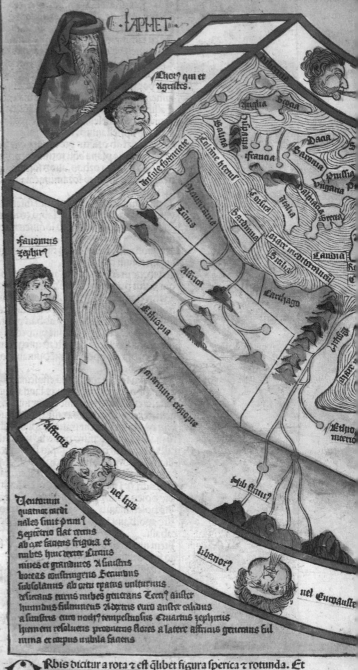

Ventorum
quatuor cardi
nales sunt prim?
Septentrio flat trans
ab are sacens frigora et
nubes huic dexter Lurius
nines et grandines A sinistris
boreas constringens Secundus
subsolanus ab ortu equalis vulturnus
siccans turis nubes generans Teen? auster
humidus fulmineus Aphricus euru auster calidus
a sinistris euru noch? tempestuosus Quartus zephirus
humen resoluens producens flores a latere affricus generans ful
mina et corpus nubila sacens

Rbis dicitur a rota & est q̄libet figura sperica & rotunda. Et
ideo mūd9 orbis dr̄. q̄ rotūd9 ē: & dr̄ orb terre vl orbisterra
rū. Dicūt āt hm vince. filij sem obtinuisse asiā. filij chā affrī
cā & filij iaphet europā. Isiō. in li. Ethy. asserit cp orbis diuisus ē in
tres partes & nō cōliter. Nā asia a meridie p orientem vsecp ad septē
trionem puenit. Europa vo a septētrione vsecp ad occidentē ptingit.
Sed affrica ad occidentem p meridiez se extendit. Sola quocp Asia

continet vnam partem nostre habitabilis. s. medietatem: alie vo ptes. s. affrica τ europa aliam medietatez
sunt sortite. Inter has autem partes ab occeano mare magnū progreditur. easch intersecat: quapropter si
in duas partes orientis τ occidentis orbem diuidas in vna erit asia in alia vo affrica τ europa. Sic autem
diuiserunt post diluuiū filij Noe: inter quos Sem cum posteritate sua asiam. Japhet europam: cham affri
cam possederunt. vt dicit glo. super Gen. x. τ super libro Paralippo. primo. Idem dicit Crisostomus Isi
dorus τ Plinius.

C These polydactyl men are described in the text of the *Nuremberg Chronicle* as being a species of men living naked and rough in the water, some of whom have six digits on their hands and feet. (Elsewhere on the overly fingered phenomenon, Pliny shares a detail in Book XI that: 'It has come down to us that the two daughters of a man of patrician family named Marcus Coranius were called the Miss Six-Fingers on this account.')

D – The Hippocentaur Schedel, author of the *Nuremberg Chronicle*, relates that, in India, there are those who live in the water and are half man, half horse. Pliny claims to have seen with his own eyes the corpse of a hippocentaur pickled in a vat of honey and transported from Egypt during the reign of Claudius Caesar.

E These are a species of woman with beards extending down to their breasts, but whose heads are bald. Again, these characters are related in the legends of Alexander, with one variation describing them hunting with dogs in the Indian mountains. 'Shun a woman with a beard as you would pestilence,' advises Pliny.

F – The Nisyti In Ethiopia, towards the west, lived men in possession of four eyes. This might have been a figurative exaggeration taken literally, for in Pliny's writings one finds the addendum: 'not that the people really have that conformation, but because they are remarkable for the unerring aim of their arrows'.

G In Europe, there were thought to be people with necks like those of cranes, and bills for mouths. These are metaphors for the good qualities of a judge, his mouth being a good distance from his heart so that it has time to consider the issue before he speaks. 'If all judges were like this', writes Pliny, 'there would be fewer bad judgments offered.'

H – The Cynocephali Dog-headed men of the mountains, who clothed themselves with the skins of wild beasts. They communicated with barking, and used their claws skilfully to hunt birds. According to a story by the Greek historian Ctesias, their population numbered around 120,000. This creature was probably based on a species of monkey, perhaps the baboon. Solinus also writes of the dog-headed Simeans of Ethiopia, who were ruled by a canine king.

I – The Arimaspi These one-eyed people lived in Scythia, 'the country to the north', in a district called Geskleithron, and were constantly at war with the Griffins over the gold that the beasts dug out of the mountains and watched over jealously. It was often the custom for people to hide money and treasure in the ground, and so the circulation of tales of such buried treasures being guarded by serpents and dragons served a purpose.

J – The Blemmyes 'In Lybia some are born headless and have mouth and eyes,' writes Schedel. The Blemmyes were a real nomadic Nubian tribal kingdom, described by Strabo as a peaceful race. They existed between 600 BC and the eighth century AD, but later became fictionalized as a legendary headless species. Shakespeare mentions them in *Othello* as cannibalistic men 'whose heads do grow beneath their shoulders', conflating them with the Anthropophagi, a man-eating race.

K – The Abarimon In what is now known as the Himalayas, in a vast country called Abarimon, the inhabitants were rumoured to be a savage race, whose feet were turned backwards but nevertheless possessed great speed, and wandered indiscriminately with wild beasts. They were unable to breathe in any other climate but their own, for which reason it was impossible to bring them before one's king. Aulus Gellius also relates this, among other wonderful tales, in the chapter, 'On the miraculous wonders of barbarous nations,' IX *c*.4. He cites, among his authorities, Aristeas and Isigonus, whom he designates as 'writers of no mean authority'.

L In Lybia lived this androgynous race: 'Some are double-sexed, the right breast male, the left one female. They are indiscriminate in their associations with one another and bear children,' writes Schedel.

M – The Sciapodes (Shadow feet) These were each owners of one very large foot and leg, while also being marvellously nimble. In the summertime, while lying on their backs, they protected themselves against the sun by the shade of their single foot. (They were also thought to live on the island of Taprobana; see relevant entry on page 220.)

N – The Straw-Drinkers 'Toward Paradise', writes Schedel, 'by the River Ganges, are people who do not eat. Their mouths are so small that they are obliged to drink through a straw. They live upon the odor of fruits and flowers. They quickly die if they encounter evil odors.'

O – The Sciritae A race of noseless, flat-faced men of short stature, who were also to be found 'toward Paradise, by the River Ganges', and whose existence is confirmed by Megasthenes.

P – The Amyctyrae (The Unsociable) Some have lower lips so large that they cover the entire face. They live on raw meat and can use their giant lower lip as an umbrella against the sun.

Q – The Panotii In Sicily lived people whose ears are so large that they cover their whole body. The ears reach to their feet and they used them as blankets to keep warm. Intensely shy, when they saw travellers they used their ears as wings with which to fly away.

R – The Satyrs In Ethiopia: '… some have horns, long noses and goat's feet; and these are spoken of throughout the legends of St Anthony', writes Schedel.

S In Ethiopia, towards the west, were people with but one foot, which was very broad. They were so fleet that they were able to pursue wild animals.

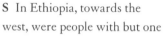

T – The Hippopodes In Scythia were people with hoofs like a horse, who, according to Pliny, lived near the Baltic.

U These people, who were only a cubit in height, had a lifespan of only eight years. They lived in the mountains of India, 'near the great sea, in a wholesome and ever verdant region'. Their wives gave birth at the age of five years, and they waged passionate warfare against cranes, their hated enemy.

PATAGONIAN GIANTS

49°18's, 67°43'w

'The existence of giants here is confirmed.'
Dr Matthew Maty, Secretary of the British Royal Society,
in a letter to the French Academy of Sciences in 1766.

Europe in the eighteenth century was dominated by the
Age of Enlightenment, a movement in which doctrines and
dogmas were challenged by science and reason – which
makes the popularity of a belief in a race of 9ft (2.7m) giants
stalking the Patagonian landscape all the more curious. In
1766, HMS *Dolphin*, captained by John 'Foul-Weather Jack'
Byron (grandfather of the poet), returned to London from a
South American expedition, bringing startling news. A new
country had been found, the inhabitants of which were all at
least 8½ft (2.6m) high. One officer of the party, Charles Clerke,
testified that he had spent two hours watching the giants being
examined and measured by Mr Byron. He stated that none of
the men was lower than 8ft (2.5m), that some even exceeded 9ft
(2.7m) and that the women were 7½–8ft (2.3–2.5m).

A detail from the map
Americae Sive Qvartae Orbis
Partis Nova Et Exactissima
Descriptio, *by Diego Gutiérrez
(1562). It indicates the 'Tierra
de Patagones' (Land of the
Big Feet) with an illustration
of two native giants clasping
bows, towering over a visiting
European.*

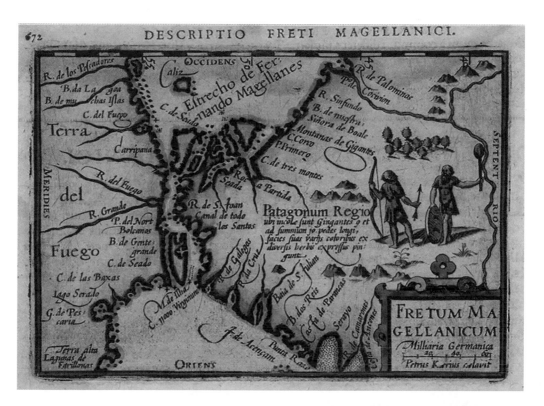

Mythical stories of giants certainly had precedent, cropping up regularly in the Old Testament, and it had also been just forty years since the publication of the journals of Jonathan Swift's Gulliver and his adventures in Brobdingnag. But, really, this does little to answer the question of why such an outlandish fallacy was widely accepted. Certainly, there were those who ridiculed the tall story, but, for many, it was confirmation of a centuries-old rumour traceable back to 1522. This was the year of Ferdinand Magellan's circumnavigation. Antonio Pigafetta, who accompanied the explorer on the journey, is commonly credited with spawning the myth in his official journal *Report on the First Voyage Around the World*, published in 1525. He describes the circumstances of a discovery following a landing at 49.5 degrees latitude:

One day we suddenly saw a naked man of giant stature on the shore of the port, dancing, singing, and throwing dust on his head … When the giant was in the captain-general's and our presence, he marveled greatly, and made signs with one finger raised upward, believing that we had come from the sky. He was so tall that we reached only to his waist, and he was well proportioned. His face was large and painted red all over, while about his eyes he was painted yellow; and he had two hearts painted on the middle of his cheeks.

Early map (c.1598) of the Strait of Magellan and Tierra del Fuego by Petrus Bertius, made for miniature atlases. The Patagonian giants stand to the right.

Magellan christened these people 'Patagoni' – 'Pata' was suspected to derive from the word for foot or shoe, and so Patagonia was translated as 'Land of the Bigfeet'. (Another theory is that Magellan was inspired by the monstrous character of Patagon in the chivalric novel *Primaleón*, published ten years before he launched his voyage.) Maps produced around this time, and for years onwards, thus bore this toponym, sometimes alternatively as 'regio gigantum' (region of the giants).

After this, there was a succession of reported giant sightings. In 1579, Francis Drake's chaplain, Francis Fletcher, wrote about coming across 'men in height and greatnes … so extraordinary that they hold no comparison with anny of the sones of men this day in the world', adding that they possessed a height of 7 foote and halfe describing the full height … of the highest of them.' In the same year, Pedro Sarmiento de Gamboa was dispatched by the viceroy of Peru to harry the English ships and chart the Strait of Magellan. His record of this mission includes a statement that the natives were 'Big People … Giants', in possession of phenomenal strength. In 1587, Sir Thomas Cavendish excitedly observed footprints 'of a gigantic race, as the measure of one of their foot marks was eighteen inches long'; while his companion Anthonie Knivet describes, in 1591, a young Patagonian 'thirteen spans tall' (about 9ft/2.7m).

The eyewitness reports then tail off. But like the dodo in a land with no predators, with a lack of contrary evidence the legend lived on to 1766, when the *Dolphin*'s findings seemed to confirm it. In 1767, an account of the Byron journey was published and became a bestseller, in large part because of its extended description of the giants and a frontispiece illustration of a member of the *Dolphin* crew offering a biscuit to an enormous Patagonian couple.

After a wary two months, the *London Chronicle* newspaper came down on the side of the *Dolphin* crew, and ran a letter drawn up by the Secretary of the Royal Society, Dr Maty, for the French Academy of Sciences: 'The existence of giants here is confirmed. Between 4 and 500 Patagonians of at least 8 or 9 feet in height, have been seen and examined by the company of one of our ships just returned from a voyage round the world; the Captain of which, who is himself 6 feet high, could hardly reach the chin of one of these men with his hand.'

Engraving based on the frontispiece illustration from A Voyage Round the World … *(1767), showing a member of John Byron's crew offering a biscuit to a Patagonian giant family.*

A. *Le Golfe d'Esperlans.*
B. *l'endroict ou nous fumes jettés contra la riue au grand danger des navires.*
C. *L'isle des Oiseaux.*
D. *L'isle des Lions.*
E. *L'isle du Roy, ou les Basteaux furent nettoyés.*
F. *C'est icy ou la Fuste se brusle.*
G. *Le lieu ou nous allames querir de l'eau avec grand' peine.*

H. *Les Sepultures des Patagons, sur le sommet des rochers, dedans lesquelles furent trouvés des os de 10, & 11, pieds.*
I. K. *Deux Lions de Mer.*
Des cerfs ayants le col si long quasi comme tout le reste du Corps.
M. *Des Austruches, qu'on en trouve a foison.*
N. *C'est une pierre, que la Nature a produict en façon de sourcils sur la cime d'une montagne.*

route

But then the tide of favour began to turn. The *Journal Encyclopedique*, one of France's most prestigious papers, published a claim that the story was a hoax designed as a distraction from the mercenary purpose of England's ships in Argentinian waters. And, in 1773, one of the most popular sets of books in the eighteenth century was published – the Admiralty's official account of the voyages undertaken by Commodore Byron, Captain Wallis, Captain Carteret and Captain Cook, in which the explorers all give the heights of the natives as just over 6ft (1.8m), puncturing the myth to the great disappointment of the readership.

It is thought that the people encountered by these Europeans were, in fact, the now-extinct, indigenous Tehuelches nomads – with an average height of 6ft (1.8m). To the Europeans (whose average height at the time was 5ft 5in (1.6m), the natives would have appeared intimidatingly large, but certainly undeserving of the label of 'giants'.

Detail from the Map of the Strait of Magellan Developed by the Schouten and Le Maire Expedition *(1616). The notation for the symbol 'H' in the lower centre translates as: 'The graves of very tall human beings, whose skeletons we found, 10 and eleven feet long.'*

PEPYS ISLAND

47°34's, 58°24'w

In November 1683, while cruising off the coast of
Guinea, the buccaneer commander Ambrose Cowley and
his restless crew of ship-snatchers spotted an anchored
Danish vessel ripe for the taking. All but a handful
of the men hid below deck to give the impression of a
harmless merchant craft, and they surprised and seized
the Dane and her forty canon with a minor loss of five
men. The pirates burnt their old ship, 'that she might tell
no tales', and in their new vessel, which they rechristened
Bachelor's Delight, set sail for the Strait of Magellan. It
was during this leg of an already eventful journey that,
in January 1684, they made a discovery that was logged
by Cowley in his journal, which was later edited and
published by William Hacke: 'In the latitude of 47°, we
saw land, the same being an island not before known.
I gave it the name of *Pepys Island*.' Hacke also includes a
drawing of Pepys Island, which features 'Admiralty Bay',
and 'Secretary's Point'.

But when one examines Cowley's original manuscript (before it
was 'Hacke'd') one finds the same entry to be substantially
different. Cowley, it turns out, never named the island 'Pepys',
nor does he describe an Admiralty Bay or Secretary's Point. It
seems Hacke embellished the account to flatter the secretary of
the Admiralty, Samuel Pepys, and trimmed a crucial reference
made by Cowley: 'We saw likewise another island by this that
night, which made me thinke them to be the Sebald de Weerts.'

The fall-out from this simple stroke of editorial air-brushing
was considerable. Pepys Island can be found on at least 111 maps
produced between 1699 and 1831 (seven of which have it drawn
but not named). The mysterious land is one of the more widely
believed phantoms, and was hunted by a veritable 'Who's Who'
of exploration. This included Dr Edmond Halley in 1698 (who
would, later, sceptically interrogate George Psalmanazar over
his claims to Formosan origins – see Formosa of George
Psalmanazar entry on page 110). Then, in the eighteenth century,
it was the object of searches by George Anson in 1740–4, John

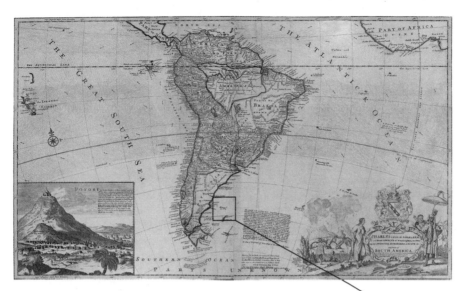

Byron in 1764, by Captain Cook during two of his famous voyages, by Louis Antoine de Bougainville and also by Jean-François de Galaup, Comte de La Pérouse, among others. Ship after ship waded through the seaweed-choked waters off the Patatagonian coast, scanning a bare horizon. They were spurred on, in 1770, by a claim of José Antonio Puig to have seen an island in Pepys's position. Conflated with 'Puig island', as well as with a contemporary phantom sought by the French known as the 'Great Island', the Pepys myth was reinvigorated.

And yet, the mystery could so easily have been solved had the entry in the Cowley journal been cross-checked with William Dampier's *A New Voyage Round the World* (1697), in which the privateer writes: 'January the 28th (1683–4) we made the *Sebald de Weerts*. They are three rocky barren Islands without any tree, only some bushes growing on them. The two Northernmost lie in 51°S, the other in 51°20'S.' Dampier was a more skilful locator than Cowley (who also wrote his journal later, from memory), and so his coordinates indicate Cowley made an error of 4 degrees. Over a century of confusion and fruitless endeavour occurred because what was overlooked – understandably since the buccaneers switched ships so often – was the fact that Dampier and Cowley were on board the *same ship*, looking at the same existent islands: the 'Sebald de Weerts'. The combination of Hacke's scissor-happy editing and Cowley's 4-degree positioning error produced an island out of thin air. The 'Sebald de Weerts', one of which Cowley mistook for a new island, are now known by another name – the Falkland Islands.

A Large Map of South America *by Herman Moll (1710), showing Pepys Island off the east coast of Patagonia.*

Following pages: W. Godson's A New and Correct Map of the World *(1702).*

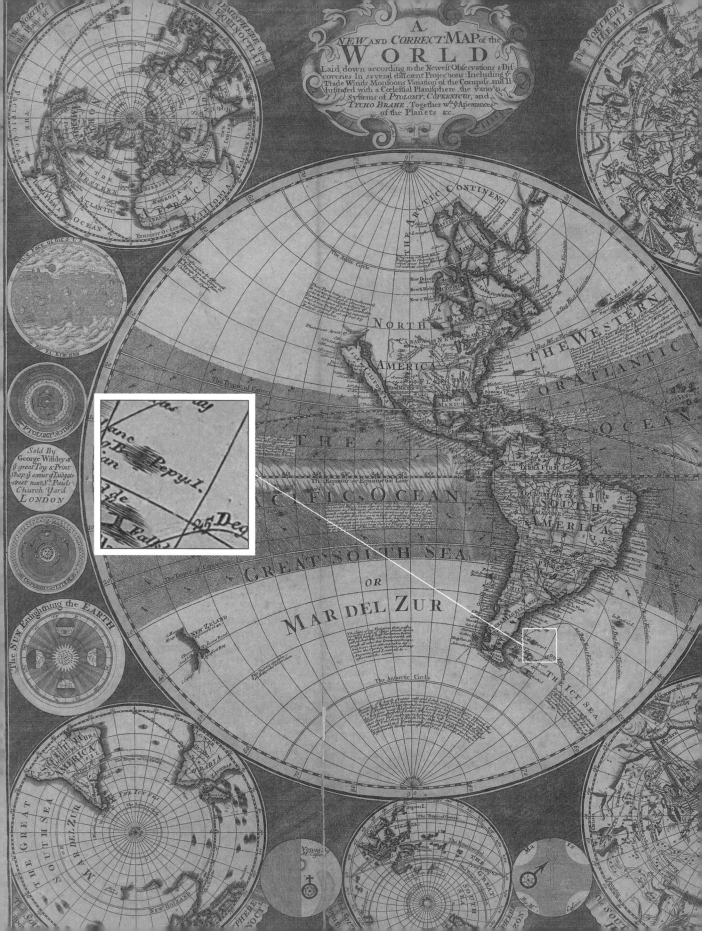

A
NEW AND CORRECT MAP of the
WORLD

Laid down according to the Newest Observations & Discoveries In several different Projections: Including y Trade Winds Monsoons Variation of the Compass, and Illustrated with a Coelestial Planisphere, the various Systems of PTOLOMY, COPERNICUS, and TYCHO BRAHE. Together w.th y Apearance of the Planets &c.

THE NORTHERN HEMISPHERE upon the Plane of the EQUINOCTIALL

THE NORTHERN HEMIS

THE PACIFIC OCEAN

NORTH AMERICA

THE WESTERN

ATLANTIC

OCEAN

AFRICA

ETHIOPIA

Ethiopic Ocean

THE FACE of the MOON By Fa KIRCHER

PTOLOMY SYSTEM

Sold By George Willdey at y great Toy & Print Shop y corner of Ludgate street next S.t Pauls Church Yard LONDON

COPERNICUS SYSTEME

The SUN Enlightning the EARTH

THE ARCTIC CONTINENT

GROENLAND

NORTH AMERICA

CANADA NEW FRANCE

CALIFORNIA

NEW MEXICO

FLORIDA

Gulf of MEXICO

The Tropic of Cancer

THE

PACIFIC OCEAN

The Equator or Equinoctial Line

TERRA FIRMA

SOUTH AMERICA

AMAZONE

BRASIL

THE WESTERN

OR ATLANTIC

OCEAN

GREAT SOUTH SEA

OR

MAR DEL ZUR

NEW ZELAND

The Tropic of Capricorn

THE ICY SEA

The Antarctic Circle

NEW HOLLAND

SOUTH AMERICA

BRASIL

MAR DEL ZUR

THE GREAT SOUTH SEA

MAR DEL ZUR

JAPON I.

The Tropic of Cancer

THE GREAT SOUTH SEA

Venus by M. Cassini

New Zeland

B. Pepys I.

Falks

25 Deg

TERRITORY OF POYAIS

15°49'N, 85°06'W

There are shameless liars, there are bold-as-brass fraudsters and then there is a level of mendacity so magnificent it is inhabited by one man alone: 'Sir' Gregor MacGregor. In 1822, South American nations such as Colombia, Chile and Peru were a new vogue in a sluggish investor's market, being lands of opportunity, offering bonds with rates of interest too profitable to pass up. And so, when the charismatic 'Cazique of Poyais' sauntered into London, resplendent in medals and honours bestowed on him by George Frederic Augustus, king of the Mosquito Coast, and waving a land grant from said monarch that endowed him his own kingdom, he was met with an almost salivary welcome. Perhaps if he had been a total stranger there might have been more wariness, but this was a man of reputation: Sir Gregor MacGregor of the clan MacGregor, great-great-nephew of Rob Roy, was famous from overseas dispatches for his service with the 'Die-Hards', the 57th Foot regiment that had fought so valiantly at the Battle of Albuera in 1811. As a soldier of fortune, he had bled for Francisco de Miranda and for Simón Bolívar against the Spanish; the man was a hero. And now here he was in London, fresh from adventure, with the glamorous Princess Josefa of Poyais on his arm, looking for investment in his inchoate nation.

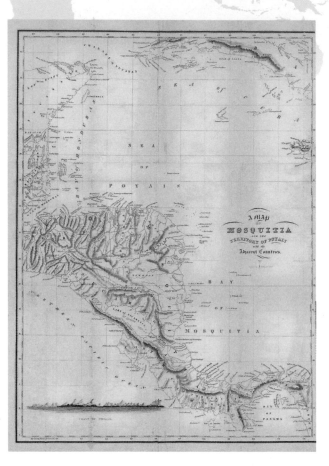

A map of Mosquitia and the territory of Poyais with the Adjacent Countries *(1822).*

And the tales he told of his new homeland! Some 8 million acres (3.2 million hectares) of abundant natural resources and exquisite beauty; rich soil crying out for skilled farming; seas alive with fish and turtles, and countryside crowded with game; rivers choked with 'native Globules of pure Gold'. A promotional guide to the region was published, *Sketch of the Mosquito Shore: Including the Territory of Poyais* (1822), featuring the utopian vista below and further details of 'many very rich Gold Mines in the Country, particularly that of Albrapoyer, which might be wrought to great benefit'. Best of all, for a modest sum you too could claim your own piece of paradise.

A dollar of the Bank of Poyais, made for Gregor MacGregor by the official printer to the Bank of Scotland.

For a mere two shillings and three pence, MacGregor told his rapt audience, 1 acre (0.4 hectares) of Poyais land would be theirs. This meant that, if you were able to scrape together just over £11, you could own a plot of 100 acres (40 hectares). Poyais was in need of skilled labour – the plentiful timber had great commercial potential; the fields could yield great bounty if worked properly. A man could live like a king for a fraction of the British cost of living. For those too 'noble' for manual labour, there were positions with prestigious titles available to the highest bidder. A city financier named Mauger was thrilled to receive the appointment of manager of the Bank of Poyais; a cobbler rushed home to tell his wife of his new role as official shoemaker to the Princess of Poyais. Families keen to secure an advantage for their young men purchased commissions in Poyais's army and navy.

One of the land grants sold to the Poyais settlers (here for a parcel of 20 acres/8 hectares).

MacGregor himself had got his start this way in the British Army at the age of sixteen, when his family purchased for him a commission as ensign in 1803, at the start of the Napoleonic Wars. Within a year he was promoted to lieutenant, and began to develop an obsession with rank and dress. He retired from the army in 1810 after an argument with a superior officer 'of a trivial nature', and it was at this point that his imagination began to take a more dominant role in his behaviour. He awarded himself the rank of colonel and the badge of a Knight of the Portuguese Order of Christ. Rejected from Edinburgh high society, in London he polished his credentials by presenting himself as 'Sir Gregor MacGregor'. He decided to head for South America, to add some New World spice to his reputation and return a hero. Arriving in Venezuela, by way of

Jamaica, he was greeted warmly by Francisco de Miranda and given a battalion to help fight the Spanish in the Venezuelan War of Independence. He then fought for Simón Bolívar when Miranda was imprisoned. Operations extended to Florida, where he devised a nascent form of what he was later to orchestrate in London, raising $160,000 by selling 'scripts' to investors representing parcels of Floridian territory. As Spanish forces closed in, he bid farewell to his men and fled to the Bahamas, never repaying the money.

Portrait of 'His Highness Gregor Cazique of Poyais'.

MacGregor was intelligent, persuasive, charisma personified, with a craving for popularity, wealth and acceptance of the elite. This was the man to whom the prospective Poyais colonists were faithfully handing their every penny. Every detail of his scheme was planned to perfection. They never stood a chance

On 10 September 1822, the *Honduras Packet* left London docks, bound for the territory of Poyais, carrying seventy excited passengers, plenty of supplies and a chest full of Poyais dollars made by the official printer to the Bank of Scotland, for which the emigrants happily traded their gold and legal tender.

Having waved off the *Honduras*, MacGregor headed to Edinburgh and Glasgow to make the same offer to the Scots. The dramatic failure of the Darien scheme in the late seventeenth century (in which the kingdom of Scotland had attempted to establish a colony on the Isthmus of Panama) had virtually bankrupted the country, and any indication of history repeating itself would have been met with extreme caution. But MacGregor was a Scotsman himself, a patriot and soldier. Unfortunately, he was also in possession of a tongue of pure silver. A second swathe of Poyais real estate was sold off, and a second passenger ship filled. Under the captaincy of Henry Crouch, the *Kennersley Castle* left the port of Leith, Scotland on 14 January 1823, carrying 200 future citizens of Poyais, eager to join the *Honduras Packet* travellers in their new home.

To their utter confusion, when the colonists arrived at their destination, they found only malarial swampland and thick vegetation with no trace of civilization. There was no Poyais, no land of plenty, no capital city. They had been fooled by a conniving fantasist. Unable to afford the journey home, they had no choice but to unload their supplies and set up camp on the shore. By April, nothing had changed. No town had been found, no help had arrived and the camp was in total despair. Disease was rife and claimed the lives of eight colonists that month. The cobbler who had been promised the role of 'Shoemaker to the Princess' gave up hope of ever seeing his family again, and shot himself in the head.

At this lowest point, a vessel appeared on the horizon – what's more, it flew a British flag. The *Mexican Eagle* from Belize had been passing nearby on a diplomatic mission when it had caught sight of the camp. The weak settlers were brought aboard and began their slow and awful journey back to London, via the hospitals of Belize. Of the 270 or so men and women who had set out for Poyais, fewer than fifty made it back to Britain. By this time MacGregor had high-tailed it to France, where he tried and failed to run the scam again. (He was foiled when the French government noticed the rush of applications for visas to a country that didn't exist.) He was eventually forced to flee to Venezuela, where he later died in 1845, never properly brought to answer for his extraordinary and terrible crime.

View of the port of Black River in the territory of Poyais.

KINGDOM OF PRESTER JOHN

Also known as Presbyter Joannes

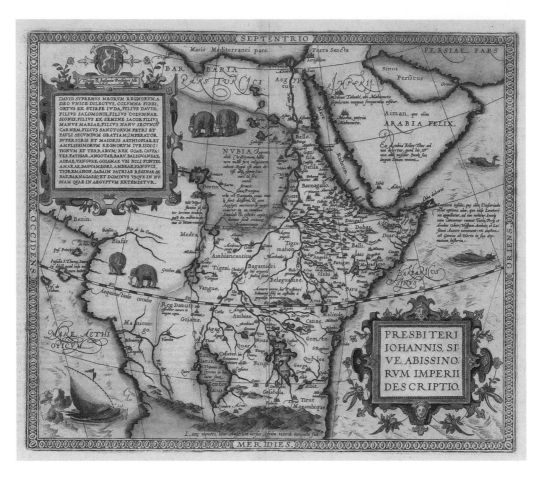

Ortelius's 1573 map of the
kingdom of Prester John in
Africa (also shows Mountains
of the Moon at the bottom; see
relevant entry on page 162).

In the latter half of the twelfth century, a mysterious
letter was copied and circulated throughout Europe to
great excitement. The message was intended for the eyes
of Manuel I Komnenos, emperor of Byzantium, bearing
greetings from, and apparently confirming the existence
of, one of the greatest figures in popular contemporary
legend – Prester John.

The priest-king John was said to be a Nestorian Christian
monarch of enormous wealth and power, a descendant of
the Three Magi (the three kings of the Bible) whose vast
realm lay somewhere in the distant and mysterious East.
The letter arrived with fortuitous timing: the Crusaders had
recently suffered a devastating defeat in Mesopotamia in

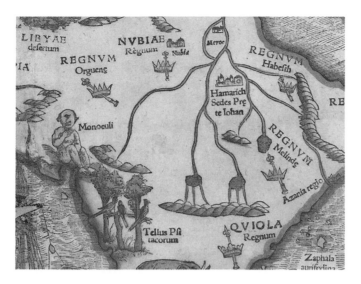

Sebastian Münster's (1550) Map of Africa, Lybia, Morland, etc. from the German edition of his Cosmographia, showing the seat of Prester John, as well as a mythical Monoculi (one-eyed man) to the west.

1144. The county of Edessa had been the first Crusader state established, and its shocking fall to Saracen forces caused huge consternation. Until the letter, Prester John was little more than a character of oral folklore; now he was a potential ally for the Crusaders. The letter proudly describes an immense and exotic domain:

If indeed you wish to know wherein consistes our great power, then believe without doubting that I, Prester John, who reign supreme, exceed in riches, virtue, and power all creatures who dwell under heaven. Seventy-two kings pay tribute to me. I am a devout Christian and everywhere protect the Christians of our empire, nourishing them with alms. We have made a vow to visit the sepulchre of our Lord with a great army, as begits the glory of our Majesty, to wage war against and chastise the enemies of the cross of Christ, and to exalt his sacred name.

Our magnificence dominates the Three Indias, and extends to Farther India, where the body of St Thomas the Apostle rests. It reaches through the desert toward the place of the rising sun, and continues through the valley of deserted Babylon close by the Tower of Babel. Seventy-two provinces obey us, a few of which are Christian provinces; and each has its own king. And all their kings are our tributaries.

In our territories are found elephants, dromedaries, and camels, and almost every kind of beast that is under heaven. Honey flows in our land, and milk everywhere abounds. In one of our territories no poison can do harm and no noisy frog croaks, no scorpions are there, and no serpents creep through the grass. No venomous reptiles can exist there of use their deadly power …

If you can count the stars of the sky and the sands of the sea, you will be able to judge thereby the vastness of our realm and our power.

The letter was a hoax, the work of an unidentified forger whose motives were equally mysterious, but the timing suggests it was designed to bolster confidence in the Crusaders' operations – for the tale of Prester John had been in existence for some time. It was first officially mentioned in European writing in the mid-twelfth century about twenty years before the letter's appearance, in the records of Otto, bishop of Freising, Germany. In the seventh book of his chronicle he writes of the meeting between Pope Eugenius III and Bishop Hugh of Jabbala in Viterbo, Italy, on 18 November 1145. Hugh had been dispatched by Prince Raymond of Antioch after the Siege of Edessa to enlist support from the pontiff. According to Otto, he then told the story of John:

'The Great Magnificence of Prester John, lord of Greater India and of Ethiopia.' Frontispiece to a popular Italian poem (undated) about Prester John by Giuliano Dati (1445–1524), Bishop of St-Léon in Calabria.

Not many years ago a certain John, a king and priest who lives in the extreme Orient, beyond Persia and Armenia, and who, like all his people, is a Christian although a Nestorian, made war on the brothers known as the Samiardi, who are the kings of the Persians and Medes, and stormed Ecbatana, the capital of their kingdom … When the aforesaid kings met him with Persian, Median, and Assyrian troops, the ensuing battle lasted for three days, since both sides were willing to die rather than flee. At last Presbyter John – for so they customarily called him – put the Persians to flight, emerging victorious after the most bloodthirsty slaughter

John was then said to have travelled with his army to aid the Church of Jerusalem, but was thwarted by the River Tigris. Unable to cross, he was forced to return home.

The search for this priest and warrior, and the kingdom he ruled with an emerald sceptre, became one of the great obsessions of the medieval age and gripped European imagination for the next five hundred years. The unifying rise of the Mongol Empire under Genghis Khan drew Western Christian emissaries such as the Franciscan explorers Giovanni da Pian del Carpine in 1245 and William of Rubruck to Asia in 1253, determined to find the lost Nestorian kingdom and secure an alliance that would save the Crusader states. As time went on and the Prester John kingdom remained undiscovered, the setting of the legend roamed around the unknown world. Toghril, the foster father of Genghis Khan, was identified with the Prester, but, when the Mongol Empire collapsed, hope was held out for finding the kingdom elsewhere and attention shifted to Africa, specifically Ethiopia, which had already the reputation of an ethereal paradise. For Europeans, 'Prester John' became the common title for the emperor of Ethiopia, or 'Third India' – despite the Ethiopians having no connection with the story.

It is this Ethiopian setting for the Prester John kingdom that Ortelius depicts in his 1573 map *A Description of the Empire of Prester John or of the Abyssinians*. In the top-left corner is illustrated the crest of Prester John, and the cartouche below it lists the monarch's lineage: 'Specially chosen by God, pillar of faith, born from the tribe of Judah, son of David, son of Salomon, son of the pillar of Zion, son from the seed of Jacob …' and so on. The map is covered in notes that relay fragments of the myths surrounding Prester John from Ortelius's time: for example, at Mount Amara (*Amara mons*), he states that here the sons of Prester John were held captive by rulers. He also shows the Niger River flowing north from Lake Niger (*Niger lacus*) and pouring underground for 60 miles (96km) before emerging in Lake Borno (*Borno lacus*); in Lake Zaire (*Zaire lacus*) can be found sirens and sea deities; and just to the east of Lake Zaire: 'Here they say the Amazons live.' The base of the map is lined by the Mountains of the Moon, 'from here southwards Africa was unknown to the ancients' (see Mountains of the Moon entry on page 162).

It wasn't until the seventeenth century that academics proved the Prester John story to have no connection with Ethiopia. It was finally accepted as a fable, albeit one likely to have had some basis in real events and characters, and is a useful example of how these fictions often played material roles in motivating early exploration, and furthering the bounds of our knowledge.

RHIPAEAN MOUNTAINS

Also known as Rhipaei Montes, Ripaei Mons

The *Orphic Argonautica* is a strange book. Once thought to be among the oldest extant poems of the Ancient Greeks, it was then discovered that the author was writing at a much later date – some time in the fourth century AD – merely imitating the style of antiquity. In a first-person account that attempts (and often fails) to mimic the Greek style, the unidentified writer reworks the story of Jason and his journey from Greece to Colchis (present-day Georgia, in the Black Sea) on a quest for the Golden Fleece, as first recorded by Appolonius of Rhodes, the head librarian of Alexandria in the second century BC. The geography of the story is one feature with which 'Pseudo-Orpheus' takes liberties – he transplants a channel, originally described as running from the Black Sea through the Balkan peninsula, further north to unexplored Russia, thereby heightening the mystery. It is through this channel that Jason sails on his return to Greece, and during these travels through the mysterious north encounters a key feature of mythical ancient geography – the Rhipaean Mountains.

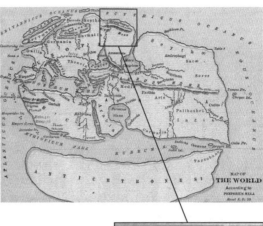

An 1890 map of the world according to Pomponius Mela, showing the Rhipaean Mountains.

In the imagined geography of the Ancient Greeks, the peaks of the Rhipaean Mountains rose up beyond the Caspian Sea to loom at the top of the known world. They were said to be the home of the mythical griffins (half-lion, half-eagle); but the myth also served a purpose: it was used to explain the origins of the Boreas, or north winds, which were said to erupt from the great caverns beneath the mountain range. Beyond the mountains lay the land of the Hyperboreans, a place of great happiness, for its inhabitants were the only ones not to be subjected to the winds' chill.

Over time, the Rhipaean Mountains were integrated into medieval European ideas of the world. Bishop Adam of Bremen, a German priest, wrote his *History of the Diocese of Hamburg* between 1073 and 1076, basing his information on interviews with – among others – King Sweyn II of Denmark, offering insights in contemporary Norse life and beliefs. In his description of Greenland, Bremen writes:

Also, there are many other islands in the wide ocean, of which Greenland is not the least; it lies farther out, opposite the mountains of Sweden, or Rhiphaean Mountains. The sailing distance to this island from Norway is said to be five to seven days, the same as to Iceland. The people there are bluish-green from the salt water, and this is what gives that region its name.

Bremen identifies the Rhipaeans with the mountains of Sweden, but, as knowledge of the area grew more thorough, the label 'Rhipaean' was found to have no connection to the area. Cartographers therefore moved it further northward, applying it to unexplored regions in the same way mapmakers filled space by drawing 'elephants for want of towns', as the Jonathan Swift quote goes. Clement Adams wrote in 1555 of the first Englishmen exploring Muscovy, and reveals a contemporary belief in the mountains, albeit dwindling, by reporting the men's failure to find them:

As regards the Riphaean hills, hoary with eternal snows, from which, as the ancients dreamed, the [river] Tanais took its rise, and the other prodigies of nature which Greece of old created out of her own imagination, our own men who lately returned home neither saw them, nor did they even so much as hear of them, although they staid in the country three months, and entered freely into conversation with the Muscovites. According to their report, the country is an open plain, and seldom rises up into hills.

It is impossible to say whether the myth of the Rhipaeans is based on an existent mountain range, but many certainly fit the bill, to the convenience of cartographers. It is the griffin, though, that in recent years has had a brighter light shed on its origins. It is thought that the mythical creature and its rumoured love of gold might well have been inspired by ancient discovery of fossilized dinosaur remains dug up on the slopes of gold-bearing hills – specifically the pentaceratops, with its bird-like beak and four-taloned feet.

Engraving of a griffin by Martin Schongauer (fifteenth century).

RUPES NIGRA

Also known as Black Cliff, Black Rock, Magnetum Insula

It was once believed that the North Pole existed in the form of a giant mountain of magnetic rock, 33 leagues (about 180km) in circumference, that stood in the centre of a polar sea – this magnetism solved the question of why compasses pointed North. Around the base of this mountain, powerful whirlpools tore the waters and drained the oceans down into the centre of the the Earth; and around this sat four separate countries.

The details of the black rock, the maelstroms and the four countries can be found on Martin Behaim's globe of 1492, but they are most famously represented by Gerardus Mercator, who first produced the idea in a small vignette on his 1569 world map. He accompanied it with a legend explaining the source of his information: a travelogue of a Franciscan friar and mathematician of fourteenth-century Oxford who, in 1360, explored the North Atlantic region on behalf of Edward III. No copy of his book *Inventio Fortunata* has survived, yet details from its pages are known thanks to a summary by Jacobus Cnoyen, entitled *Itinerarium*. What we learn from that text is that *Inventio Fortunata* was an extraordinary work of imagination. Mercator repeats the friar's description of the North Pole, writing on the map: 'He averred that the waters of these 4 arms of the sea were drawn towards the abyss with such violence that no wind is strong enough to bring vessels back again once they have entered; the wind there is, however, never sufficient to turn the arms of a corn mill.'

This seemed to correlate with a similar scene described by the historian Gerald of Wales (*c.*1146–*c.*1223) on the marvels of Ireland, who writes:

Not far from the isles (Hebrides, Iceland, etc.) towards the North there is a monstrous gulf in the sea towards which from all sides the billows of the sea coming from remote parts converge and run together as though brought there by a conduit; pouring into these mysterious abysses of nature, they are as though devoured thereby and, should it happen that a vessel pass there, it is seized and drawn away with such powerful violence of the waves that this hungry force immediately swallows it up never to appear again.

Mercator expanded his Arctic image with a larger dedicated map in his atlas of 1595, featuring at the centre the magnetic mountain island 'Rupes Nigra', which he described in a 1577 letter to John Dee as 'black and glistening' and 'high as the clouds', surrounded by the raging sea. Other mythical features worthy of particular note on this map of Mercator are the inclusions of Groclant, southwest of the pole (see relevant entry on page 128); Frisland (see Phantom Lands of the *Zeno Map* entry on page 240) in the top-left corner; a separate magnetic rock north of the pole at the mouth of the Strait of Anian (see relevant entry on page 12); and the inscription on the outcrop of land directly southeast of the pole, which translates as: 'Here live pygmies whose length is no more than 4 feet.'

Athanasius Kircher incorporates the belief in his *Mundus Subterraneus* (1665), suggesting the system of the Earth's water flow might be comparable to that of the human body. His theory was that the seas ran through the Bering Strait and into the fabled North Pole vortex, where, via 'unknown recesses and tortuous channels', the waters poured down through Earth to burst out at the South Pole.

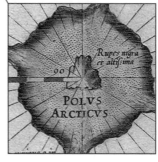

Mercator's Arctic projection (this of 1606) was the first specifically of the North Pole. In the centre can be found 'Rupes nigra et altissima', the mythical magnetic black rock.

ST BRENDAN'S ISLAND

28°32'N, 23°14'W

*Also known as the Isle of the Blessed, San Borondón,
the Promised Land of the Saints*

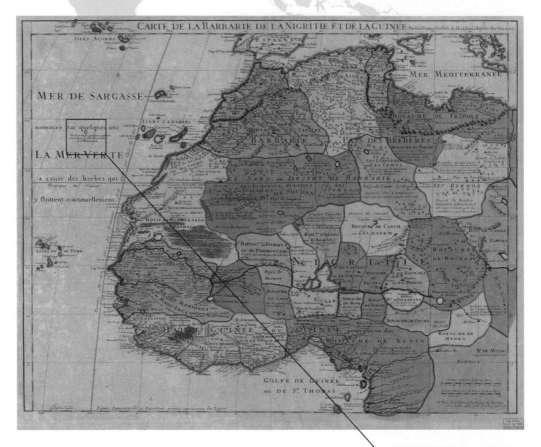

The Voyage of St Brendan is the most famous of the five immrama (sea tales) of the early Irish. Though relayed with increasingly mythological flourish over the years, the story of St Brendan's mission to discover the 'Promised Land of the Saints' contains elements of geographical accuracy that suggests the adventure might really have taken place in some form – the references to a frozen sea, and encounters with an iceberg and a volcano being of particular interest.

According to the story, in around the sixth century, St Brendan was inspired to find the Promised Land of the Saints by St Barrind, who claimed to have visited an island paradise with his disciple Mernóc. St Brendan decides to see the land for

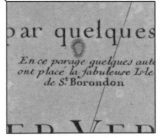

On Guillaume de l'Isle's Carte de la Barbarie de la Nigritie et de La Guinee *(1707), just west of the Canary Isles, the French cartographer draws the island with the note: 'In this vicinity several authors have placed the fabulous Isle of St Borondon.'*

himself, and assembles a group of fourteen monks to go with
him. In preparation, the men fast for three-day intervals for
forty days, then visit a nearby island to seek the blessing of
St Enda for their mission. The monks then build a currach
from wattle, timber and ox hides tanned in oak bark and
softened with butter. They fit the small boat with oars and a
sail, and enough provisions for forty days' sailing. At this point,
three latecomers arrive to join the mission, and unwittingly
place a curse on the journey by altering the sacred number of
participants (a detail shared by another immrama, 'Voyage of
Mael Duin').

The monks set sail, and after an exhausting forty-three days
at sea, with their rations almost depleted, land on a deserted
island. There they discover a great hall with food mysteriously
laid out, empty but for a dog and an Ethiopian devil. One of
the latecomers then admits to having stolen from the island. In
response, St Brendan performs a ritual and the Ethiopian devil
is exorcised from the man, who later dies and is buried by his
comrades. Convinced that this island is not the one described
by St Barrind, the monks set sail again, and land at a series of

*St Brendan celebrating Mass
on the back of a whale, from
Caspar Plautius's* Novo Typis
Transacta Navigatio …
(1621).

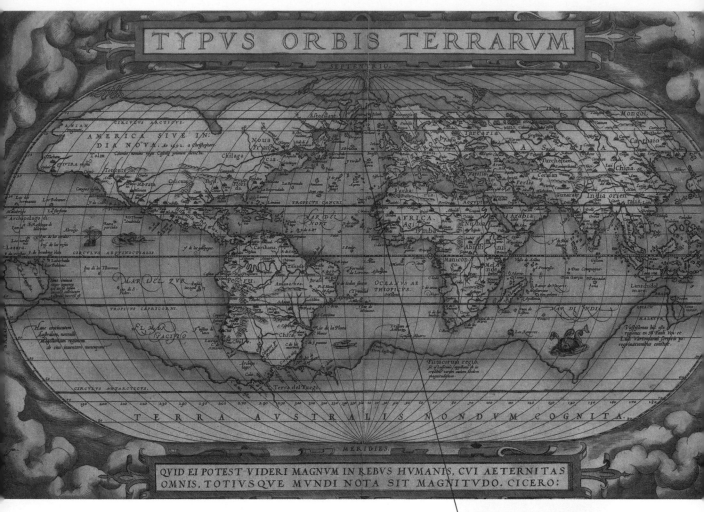

islands: on the first they are met by a young boy who presents
them with bread and water. On the next, they encounter flocks
of sheep taller than cattle (on which the monks dine heartily).
On another island, called Jasconius, they celebrate Easter Mass
only to discover that they are, in fact, standing on the back
of a giant whale – the enormous creature slowly begins to
dive below the surface, and the men escape just in time. After
that they find the 'Paradise of Birds', on which flocks of birds
sing psalms in praise of God. This is followed by an island of
magic loaves, agelessness and total silence, where they find
a monastery of the monks of Ailbe, a brotherhood sustained
entirely by their faith in God for more than eighty years.

After exploring a last island, on which the water from a well
sends the men into a deep sleep, the monks launch again and
become caught in a 'coagulated' (frozen) sea. They eventually
break free of the ice and return to visit again the islands of

On Ortelius's World Map *(1570),
'S. Brādain' can be found just
below the fictitious Drogeo in the
North Atlantic.*

sheep, the whale Jasconius and the Paradise of Birds. While on the last, a bird warns the holy men that they must continue this year-long cycle for the next seven years if they wish to be worthy of the Island of Paradise.

Out at sea once more, the monks' vessel is rocked by a giant sea creature intent on devouring them, but God whirls the water around them for protection and another equally giant creature arrives in the nick of time to kill the aggressor. The men feast on the flesh of the dead monster and then continue on their way to an eerily flat island, the surface of which is at sea level. They are greeted by three choirs of Anchorite monks who give them white and purple fruit, one piece of which can sustain a man for twelve days.

After surviving an attack by a griffin, they sail a transparent sea and pass a 'silver pillar of bright crystal', which is thought to be a reference to an iceberg. They push on, soon reaching the 'Island of the Smiths', where they hear terrifying sounds and see smoke rise from a boiling sea (this presumably is a volcano). They escape, only to find another volcano, a mountain where 'great demons threw down lumps of fiery slag from an island with rivers of gold fire'. Onwards again, this time to find two islands each occupied by a hermit – the first man turns out to be Judas living in exile, whom St Brendan protects from demons; the second is St Paul, who has lived for 140 years off fish given to him by a friendly otter. After another visit to the islands of birds, sheep and Jasconius, the monks' vessel becomes wrapped in dense fog and they finally reach the island for which they have spent seven years searching. They stay there briefly, collecting fresh fruit and precious stones, before making the voyage home.

This was a popular tale that was widely spread. The 'St Brendan's Isle' is featured on the Ebstorf world map of 1235, as well as the map made by Paolo Toscanelli del Pozzo for the king of Portugal. The cartographic position of the island constantly shifted: initially, the island was associated with the Canary Islands, but was gradually shunted further into the Atlantic. Ortelius's map of 1570 puts it several thousand miles north of the Azores, near the coast of Newfoundland, where it remained on maps well into the seventeenth century.

SANDY ISLAND, NEW CALEDONIA

19°13's, 159°55'E

Also known as Île de Sable

The revelation that Sandy Island was an unremarkable patch of ocean came in November 2012. For more than one hundred years, it had been charted with specific coordinates in the Coral Sea, sat between the Chesterfield Islands and the Nereus Reef, northeast of Australia and west of Caledonia.

UK Hydrographic Office chart of the Pacific Ocean by R. C. Carrington, noting that Sandy Island was observed by the ship Velocity in 1876.

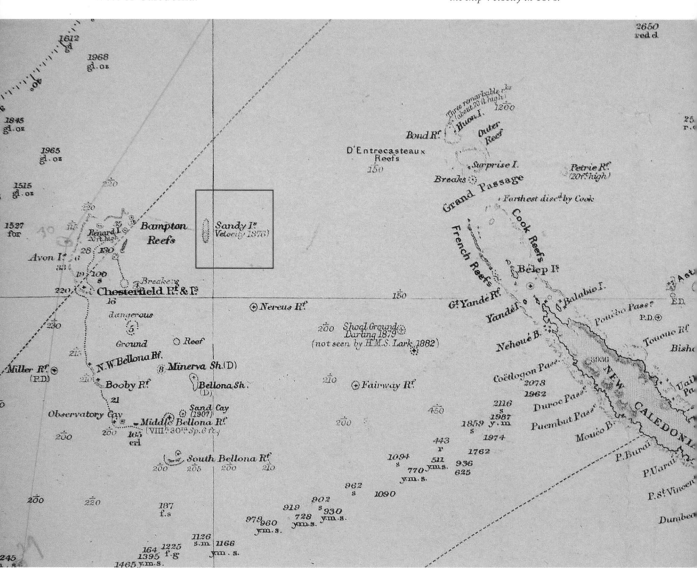

Then, in 2012, it happened that a team of Australian marine scientists aboard RV *Southern Surveyor* were conducting a study of plate tectonics in the area, and noticed the island on their data sets. They decided to include it in their route, and when they reached the location, instead of the expected sliver of sand and palm tree, they were instead puzzled to find nothing at all, with the water depth measuring 4265ft (1300m). After further checking, they found that, although clearly shown on Google Maps, it did not appear on the navigational charts of the ship. The missing island was initially attributed to a technical error in the data sets, including that used by Google; however, the case of Sandy Island is representative of the problem that occasionally arises from the fact that modern digital maps are drawn from a combination of data from satellite imagery and some of the oldest maps of the British Admiralty. In this instance, it's entirely possible the phantom can be traced back to 1774, when Captain James Cook recorded a 'Sandy Island' at a location 260 miles (480km) further east, with a four-degree difference in longitude. The whaling ship *Velocity* sighted the island, in 1876, nearer the modern coordinates, and their findings of 'heavy breakers' and 'Sandy Islets' were placed on various sea charts of the late nineteenth century, including an 1895 British Admiralty chart; they then appeared in an Australian maritime directory for 1879. Their position was written as running north and south 'along the meridian 159°57'E' and 'between lat 19°7'S and 19°20'S'.

On French hydrographical charts, Sandy Island began to be cleared from 1979, and in 2000 it was pointed out that the island was also not to be found in the 1999 *Times Atlas of the World*. Today, the island's location can still be spotted with a search on Google Maps, though it is now accompanied by a caption explaining its recent undiscovery.

'It's unlikely someone made this island up,' said Danny Dorling, president of the Society of Cartographers, to reporters in 2012. 'It's more likely that they found one and put it in the wrong location. I wouldn't be surprised if the island does actually exist, somewhere nearby.'

SANNIKOV ISLAND

78°53'N, 147°20'E

Also known as Sannikow Land, Zemlya Sannikova

In 1810, a Russian cartographic expedition led by the explorer Matvei Gedenschtrom set off for the New Siberian Islands, where average winter temperatures lie around -27°F (-32.9°C). During this punishing journey one of the party, Iakov Sannikov, sighted an unknown land in the Arctic Ocean, north of Ostrov Kotel'nyy (situated off Russia's northern coast,

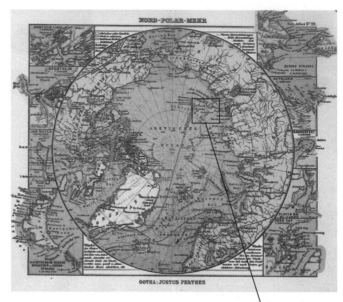

between the Laptev Sea and the East Siberian Sea). Sannikov was a reliable geographer – he had previously discovered and charted nearby Stolbovoy Island in 1800 and Faddeyevsky Island in 1805 – and this new land claim, described by Sannikov as emanating a 'bluish fog', was backed up by the team leader, Gedenschtrom. The discovery was christened 'Sannikov Land', and would not be seen again for seventy-five years. In 1886, the German geologist and Arctic explorer Baron Eduard Vasil'yevich Toll was conducting surveys of the New Siberian Islands. The expedition, led by Aleksandr Bunge, met with huge success when the men dug through the fossil-ice of the south coast of Lyakhovskiy Island and discovered a wealth of neatly preserved remains of mammoth, rhinoceros, antelopes and even a sabre-tooth tiger, as well as a giant 20ft- (6m)-high specimen of the shrub *Alnus fruticosa*. To add to the excitement, during this expedition Toll also reported seeing an unknown land north of Kotelny, which he thought might be the 'Zemlya Sannikova'.

'Sannikow ld.' is placed just outside the Arctic Circle on this German map from 1906.

In 1900, Toll was sent back to the region by the St Petersburg Academy of Sciences to lead an expedition over the Laptev Sea of the Russian Arctic in the vessel *Zarya* in search of Sannikov Land. The crew was made up of decorated naval officers and experienced seamen, as well as the astronomer F. G. Zeiberg and two men named Gorokhov and Protodyaknonov, who are described in records as 'entrepreneurs'. The *Zarya* launched in June 1900, but by the spring of 1901 had only made it as far as the mouth of the Yenesey River, where she stopped to resupply. In August, Toll and his crew became only the fourth expedition to successfully round Cape Chelyuskin, the northernmost point of mainland Russia. They then began their search for Sannikov Land, heading for Bennett Island in the east Siberian Sea. They found nothing, and so drew back to the New Siberian Islands to see out the winter on Ostrov Kotel'nyy. The *Zarya* became trapped by floating pack ice, and so, in July 1902, Toll and several of his men courageously decided to leave the ship and strike out by sledge and kayak back across the ice for Bennett Island, crossing 93 miles (150km) in a northerly direction. When the *Zarya* was finally freed in August, the crew set out to rescue Toll and his group, but was unable to make it through the ice and had to divert to the mouth of the River Lena, where the men were forced to return to St Petersburg. With no sign of the *Zarya* on the horizon, Toll and his team decided to attempt the trek back to Ostrov Kotel'nyy themselves, despite being weak and short on supplies. They were never heard from again. Two search teams sent in 1903 managed to recover Toll's diary (which was later published by his wife in 1909), but found no other trace.

The crew of the Zarya, *c.1901, with Baron Eduard Toll in the centre.*

Now freighted with tragedy, Sannikov Land was sought a final time in 1936 by the Soviet icebreaker *Sadko*, but in 1937 it was concluded that it no longer existed, if indeed it ever had, and it was dropped from maps. Possibly it was an island that once existed but was destroyed by phenomenally fierce erosion, or perhaps what was glimpsed was a permafrost-saturated shoal that had simply sunk (a common phenomenon in the area). The most likely explanation, though, is that Sannikov and Toll were the victims of the same kind of complex mirage that tricked Donald MacMillan into his mad pursuit of Crocker Land (see entry on page 70): a Fata Morgana, which presents itself as a distant strip of cape lying tantalizingly close, and yet always just out of reach.

SATANAZES

40°13'N, 48°25'W

Also known as Isle of Devils, Satanaxio, Santanzes, Satanagio, La Man Satanaxio, Salvatga, Salirosa

On the 1424 portolan chart of the Venetian cartographer Zuane (or Giovanni) Pizzigano that bears the mythical island of Antillia (see relevant entry on page 18), another large and equally mysterious island is shown 60 leagues (290km) to the north, named 'Satanazes'. This is the first depiction of the island also commonly referred to as the 'Isle of Devils', one of several demonic islands imagined in the Atlantic, situated due west of the Azores and Portugal. Pizzigano's chart shows five cities marked on Satanazes, called Aralia, Ysa, Nar, Con and Ymana. These change frequently on later maps: Grazioso Benincasa's atlas of 1463, for example, lists six: Araialis, Cansillia, Duchal, Jmada, Nam and Saluaga. The island appears on many of the significant maps of the fifteenth century, including those of Battista Beccario (1435, as 'Satanagio'), Pedro Roselli (1480, as 'Salvatga') and

The Pizzigano portolan chart of 1424 was the first to carry Satanazes (Isle of Devils), depicted as a rectangular island to the left.

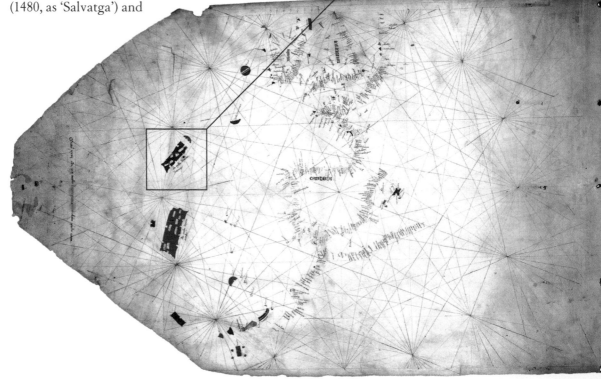

the Laon globe (1493, as 'Salirosa'). After the voyages of Christopher Columbus in the 1490s, however, Satanazes disappears from cartographic record, possibly relocating as the 'Isle of Demons', another phantom island once believed to exist on Quirpon Island, Newfoundland in Canada (see Isle of Demons entry on page 84).

It is thought, by some historians, that the 'devils' of Satanazes could refer to indigenous populations encountered by the Norsemen, who also labelled these people Skraelings, and that Pizzigano might, therefore, have gathered the information for his portolan chart from, among other sources, the Norse sagas. Just as with Antillia, it is not known which country Satanazes might represent – North America has been proposed as one contender. Interestingly, an alternative label for Satanazes on some maps, such as Andrea Bianco's 1436 chart, is 'Ya de la man santanaxio', a translation of which, suggested by Vicenzo Formaleoni in 1783, is 'Isle of the Hand of Satan'.

The legend of an isle that is home to the Hand of Satan has a long history in sea-tale mythology. In *Tales of the Enchanted Islands of the Atlantic* (1898), Thomas Wentworth Higginson writes an especially vivid account:

Fogs gathered quickly, so that they could scarcely see the companion boat, and the Spanish fishermen called out to them, 'Garda da la Man do Satanaxio!' ('Look out for Satan's hand!') As they cried, the fog became denser yet, and when it once parted for a moment, something that lifted itself high above them, like a gigantic hand, showed itself an instant, and then descended with a crushing grasp upon the boat of the Spanish fishermen, breaking it to pieces, and dragging some of the men below the water, while others, escaping, swam through the ice-cold waves, and were with difficulty taken on board the coracle.

Nils Nordenskiöld, the nineteenth-century Arctic explorer, offers an alternative theory: he suggests that the name might derive from 'Santanagio', the Basque form for St Anastasius; and that it could even be evidence of Basque exploration of far Atlantic – and possibly even North American – waters, before the time of Columbus.

'A demon hand sometimes uprose from the islet and plucked away men and even whole boats, which, when once grasped, usually by night, were never seen again, but perished helplessly' (T. W. Higginson's Tales of the Enchanted Islands of the Atlantic, *1899*).

'Saxenburg Isle', taken from Herman Moll's 1710 map of South America.

Supposedly first sighted by a Dutch merchant named J. Lindeman of Monnikendam at sunset on 23 August 1670, this island off the east coast of South America was described as having a narrow peak or column rising near the centre of the flat island, giving it the appearance of a witch's hat. More than 100 years went by before any further mention was made of it, until Captain Flinders sought it in 1801, with no luck. 'His precautions were such', writes John Purdy in his *Memoir* (1822), 'as to leave no doubt of the nonexistence of the island within the limits here mentioned.' The famously meticulous Captain James Horsburgh writes in *Directions for Sailing to and from The East Indies …* (1809) that he had launched a search for Saxenburgh on two separate occasions, and had concluded its existence doubtful: 'About this place I have seen clouds, exactly like land, remain stationary at the horizon for a great length of time; a superficial observer might have taken a view, and asserted that it was an island. These doubtful islands and dangers are very perplexing to navigators.' He attests to this last point with a story of two ships reaching Saxenburgh's coordinates on their way to India: 'A heavy cloud near them was taken for the island, and they conceived the swell or current was setting them fast upon it, the weather being unsettled and calm at the time; this induced them to get out the boats to tow the ship's head off shore, till the island vanished with the dawn of day.'

Nevertheless, one finds an intriguing footnote in Matthew Flinders' account, *A Voyage to Terra Australis* (1814):

At the Cape of Good Hope, in 1810, His Excellency the Earl of Caledon favoured me with the following extract from the log book of the sloop Columbus, – Long, Master; returning to the Cape from the Coast of Brasil.

September 22d, 1809, at five p.m. saw the island of Saxonberg, bearing E.S.E. first about 4.5 leagues distant; clear weather. Steered for the said island, and found it to be in the latitude of 30°18'S., longitude 28°20'W., or thereabout.

The island of Saxonberg is about four leagues in length, N.W. and S.E., and about two miles and a half in breadth. The N.W. end is a high bluff of about seventy feet, perpendicular form, and runs along to the S.E. about eight miles. You will see trees at about a mile and a half distance, and a sandy beach.

And, indeed, Captain Galloway of the American ship *Fanny*, outward bound to China, had managed to sail for four hours in sight of Saxenburgh in 1804, and had echoed Lindeman's original description of it possessing a central peak, as well as a bluff to one extremity; though he marks the island two degrees further east. In 1816, Captain Head of the *True Briton* also claimed to have laid eyes upon it for six hours; but still no man had yet managed to set foot on its shores.

Despite this intangibility, Saxenburgh was embraced by naturalists and geographers. Nathaniel Dwight's 1817 *A System of Universal Geography, for Common Schools*, tested British schoolchildren with the following: 'Question: Where are Gough's Island, Diego, Tristan de Cunha and Saxemberg? Answer: W. of the Cape of Good Hope and nearly its latitude.'

Alexander Beatson, governor of St Helena island in the South Atlantic, claimed in 1816: 'I have in my possession a sketch of the Island of Saxemberg, upon which some trees are represented; of what sort I am not informed', and postulated the land mass might once have been 'united' with nearby Gough's Island and Tristan da Cunha.

Saxenburgh Island was also sought by Benjamin Morrell, who already possessed an impressive phantom roster of Byers's Island and Morrell's Island (see Lands of Benjamin Morrell entry on page 166). In *A Narrative of Four Voyages to the South Sea*, Morrell writes:

August 18th … *I now determined to sight the island of Saxenburgh, if such an island really existed within any reasonable distance of the spot in which it is said to be situated … We were roused by the cheering cry from the mast-head of 'Land, ho! Land, ho! About six points off the starboard bow … We now had the wind from west-by-south, which permitted us to haul up for it; but after running in that direction about four hours, at the rate of eight miles an hour, our tantalizing land took a sudden start, and rose about ten degrees above the horizon. Convinced that we could never come up to it in the ordinary course of navigation, we now tacked and stood to the northward. We had likewise seen land the day before, at 4, P.M., exactly in our wake, which appeared to be about twenty miles distant.*

In Morrell's opinion, those who believed they had sighted Saxenburgh Island had probably been tricked by large cloud formations, 'exactly like land in appearance, [they] will sometimes remain stationary at the horizon in this part of the ocean, for a great length of time, and are easily mistaken for distant islands'.

Indeed this was now the consensus at the time, and Saxenburgh was no longer included on maps. But the strangest part of this episode is that, in 1965, Morrell's mission to find Saxenburgh was shown to be completely fictitious. This was when the logbook of John W. Keeler, who had accompanied Morrell on two of his voyages, came into the possession of the G. W. Blunt Library of Connecticut. Keeler's meticulous records showed that the course of Morrell's schooner *Atlantic* never went anywhere near the positions at which Morrell claimed to have searched for Saxenburgh – he'd concocted the episode to spice up an otherwise unremarkable journey.

SEA OF THE WEST

37°53'N, 118°45'W   *Also known as Mer de l'Ouest, Baye de l'Ouest*

This extraordinary map, *Carte des Nouvelles Decouvertes* (1750) by Joseph-Nicholas de l'Isle and Philippe Buache, presents perhaps the most spectacular artistic embodiment of the fanatical search for the Northwest Passage – La Mer de l'Ouest.

As usual, the strange misbelief was not plucked from the air – in fact, this was a theory centuries in the making. The first explorers of the North American coast were teased by inlets, bays and straits that all appeared to be potential channels through the country to the ocean on the other side. In 1524, a great theoretical eastern inlet was created when Giovanni Verrazano sailed in a southerly direction along the east coast of North America and came to the thin Outer Banks of North Carolina. Looking past them, he spotted a glittering sea on the other side, and excitedly assumed he had found a short way to the Pacific:

From the ship was seen the oriental sea between the west and north. Which is the one, without a doubt, which goes about the extremity of India, China and Cathay. We navigated along the said isthmus with the continual hope of finding some strait or true promontory at which the land would end toward the north in order to be able to penetrate to those blessed shores of Cathay ...

Sixteenth-century European maps began to represent the idea that the Atlantic and Pacific were divided by only a pencil-thin stretch of North America, which is why we find the country presented by Sebastian Münster in such a twisted form, showing a huge eastern inlet. Then, in 1625, Samuel Purchas published *Purchas His Pilgrimes ...,* a history of exploration that included Michael Lok's account of Juan de Fuca. The almost certainly invented Greek sailor (who featured in the story of the mythical Strait of Anian) journeyed up a vast strait at the northernmost point of America's west coast, around which he enjoyed 'sayling therein more than twenty days'. Hugely influential, this trusted information was employed by cartographers who explained the twenty-day voyage with the logical speculation of a large western inlet.

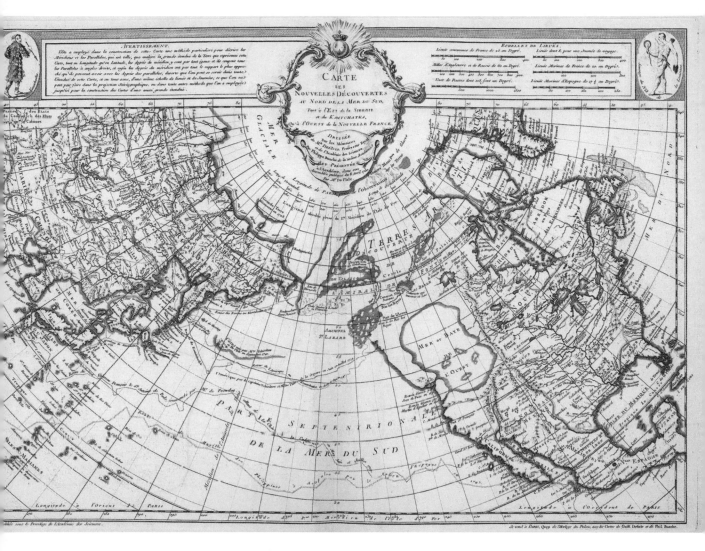

Carte des Nouvelles
Decouvertes, *a map by Buache
and de l'Isle from 1750, showing
La Mer de l'Ouest, a vast inland
sea in North America.*

As European settlement of North America expanded, the
English and French grew more desperate for a thoroughfare
trade route to compete with the Spaniards' lucrative East
Indies trafficking. Those who mounted expeditions through
the American interior relied on information from native
guides, and, as we have seen from Baron de Lahontan's error-
ridden journey tracing his Great River of the West (see relevant
entry on page 126), these explorers were not provided with the
most reliable of intelligence. Nevertheless, the cartographer
Guillaume de l'Isle combined the now-established rumour of
a Western Sea with Lahontan's accounts of a great salt lake
(as described by Lahontan's quoted source, the Tuhuglauks)
to feature the body of water on his lauded 1703 map of
Canada. In 1717, the French Navy Council and Louis XV

Tabula nouarum insularum, quas diuersis respectibus Occidentales & Indianas uocant.

expressed interest in La Mer de l'Ouest and, in 1729, the Jesuit missionary Pierre de Charlevoix was dispatched to investigate, with no success.

For the next thirty years, no evidence of such a sea was found and the idea was dropped from discussion, until Joseph-Nicholas de l'Isle (son of Guillaume) and Philippe Buache, both highly respected and assiduous geographers, resurrected the idea in 1750. To fill in the gaps, they also drew upon the famous 'letter of Admiral Bartholomew De Fonte'. Printed in 1708 by the London magazine *Memoirs for the Curious*, this mysterious document purported to be written by the Spanish

Sebastian Münster's Tabula novarum insularum, quas Diversis respectibus Occidentales & Indianas uocant *(1554) was the first printed map of the American continent. North America is shown bent over backwards to accommodate Verrazano's description.*

sailor De Fonte recounting a voyage along an eastward channel through North America, in which he encounters a ship coming the other way, from Boston. The letter, which appeared to prove secret Spanish knowledge of a northwest passage in a temperate parallel, and which was an explosively effective catalyst in reigniting the search for the route, is now considered a fake, most likely written by the editor of *Memoirs* or created as a motivator for public support. On these most rotten of foundations, Buache and de l'Isle built their theories presented on the map shown on page 217. The mythical Sea of the West would float on maps until 1786, when Jean-François de Galaup, comte de Lapérouse, methodically recorded the coasts between Mount St Ellias and Monterey, and dispelled the myth once and for all.

Incidentally, this was not the only spectral sea of Philippe Buache – his conjectural map of the Antarctic, *Carte des Terres Australes*, depicts his belief in a great body of water, the Mer Glaciale, lying at the centre of the Antarctic. This he considered the most likely explanation for the icebergs seen on the region's coast by the French explorer Jean-Baptiste Charles Bouvet de Lozier in 1738. To be fair to Buache, though, his analysis of the latest journals, reports and other data brought back by travellers also occasionally led to remarkably accurate deductions – he was confident, for example, in the existence of both Alaska and the Bering Strait before they were officially confirmed.

Buache's other conjectural sea, at the centre of the Antarctic. From Carte des Terres Australes, *1754.*

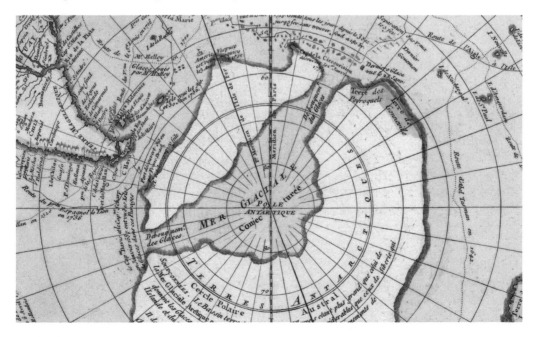

TAPROBANA

7°30'N, 80°44'E

Also known as Taproban, Taprobane

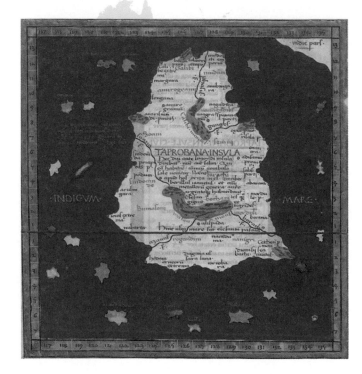

Nicolaus Germanus's
Taprobana Insula *from*
Cosmographia Claudii
Ptolomaei Alexandrini *(1467).*

Just as the Cassiterides, with its resources of tin, had its location kept a secret by merchants (see relevant entry on page 68), so too did Taprobana. The large island, abundant in cinnamon, pepper and other valuable spices, lay somewhere in the Indian Ocean and was first described in 290 BC, by the Greek explorer Megasthenes, who wrote that it was divided by a river and was more productive of pearls and gold of great size than India.

Pliny refers to it as the distant counter-land of the 'Antichthons', where everything was upside-down. The sea between the island and India was full of shallows barely 'six paces' in depth, and in other places so deep that anchors failed to reach the sea floor. The local ships were built with prows at both ends so as not to have to turn in narrow channels, which the Taprobane mariners navigated not through observations of the stars, but by birds, which they released and followed. Pliny then provides specific details from the reign of Claudius, with the story of a unnamed freedman working for a Red Sea tax collector named Annius Plocanus. The man was carried away by gales from the north, while sailing around Arabia, and after fifteen days found himself arriving at Hippuri, a Taprobanian port. He was warmly received by the king, who 'dresses like Father Bacchus', and in six months had learnt the language and was able to answer questions about Rome, which resulted in four Taprobanian ambassadors being sent there to develop relations. From these men it was learnt that Taprobana was a nation of 500 towns, the most magnificent of which was Palaesimundus, home to 200,000 including the royal family. At the bottom of their seas, which were bright green, grew forests of trees that often broke the rudders of ships; and the people entertained themselves by hunting elephant, tiger and turtles so vast that 'beneath the shells … whole families can be housed'.

Many legends swirl around Taprobana: the compiler of the fourteenth-century *The Travels of Sir John Mandeville* (whose identity is disputed) states that the island was within the kingdom of Prester John (see relevant entry on page 194) that winter and summer occurred there twice annually, and (in Chapter 33) that its mountains of pure gold were defended by enormous man-eating ants. The last detail was probably inspired by the writings of Pomponius Mela of AD 43, who described ants as large as mastiffs: 'In the isle also of this

Ptolemy's Taprobana, by Michel Servet, published in the 1535 edition of Cosmographia Claudii Ptolomaei Alexandrini.

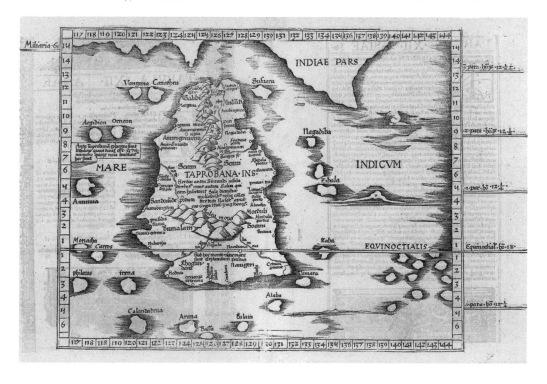

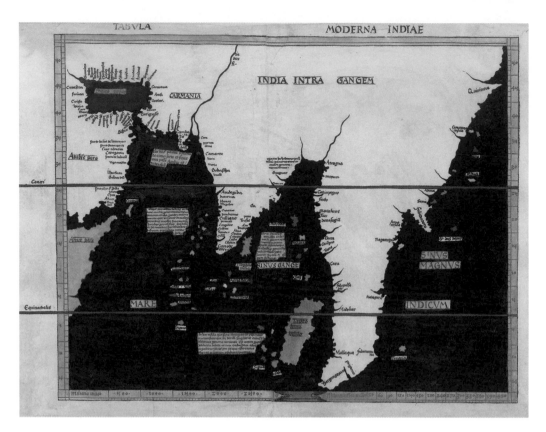

Taprobane be great hills of gold, that pismires [ants] keep full diligently. And they find the pured gold, and cast away the unpured. And these pismires be great as hounds, so that no man dare come to those hills for the pismires would assail them and devour them anon …'

To get at the gold, the author of *The Travels …* writes that the Taprobanians chased the giant ants away by charging them down on camels, horses and other beasts. They also used a more subtle tactic, hanging vessels on horses and guiding them to graze on the gold hills. When the ants saw the empty containers carried by the horses, they ran towards them and filled them with gold, for 'they have this kind that they let nothing be empty among them, but anon they fill it be it what manner of thing that it be …'

Taprobana was also said to have been populated by the mythical race known as the Sciapodes, men with one giant foot who used their limb for shade from the noon sun as they lay on their backs. The Sciapodes (or Monopodes) are mentioned by Aristophanes in his play *The Birds* (414 BC) and by Pliny in his *Natural History*, who reports travellers' accounts of encountering

Waldseemüller's modern map of the Indian Ocean, India, southeast Asia and adjoining regions (including 'Taprobana Insula'), from the 1513 edition of his Geographia.

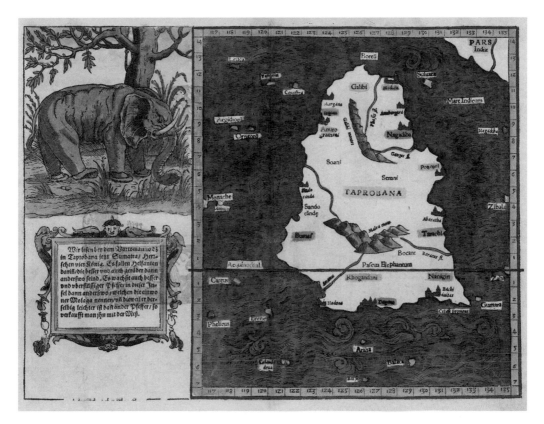

the creatures in the Indian region; and the medieval *Hereford Mappa Mundi* features an illustration of the Sciapod in the Indian area. Also on the subject of Taprobana's inhabitants, in *The Most Famous Islands in The World* (1590), Tommaso Porcacchi repeats the description by Diodorus Siculus of people with forked tongues: 'Double down to the root & divided; with one side they speak to one person, and with the other they speak to someone else.' Porcacchi then apologises to his readers for being unable to find the land's location.

Though it has not been conclusively established which island can lay claim to being the Taprobana of old, Sumatra is one of the most popular suggestions, which is how it was identified by Niccolò de' Conti in the fifteenth century; Sebastian Münster's map of the island has also been pointed to, with its German title *Sumatra Ein Grosser Insel* (*Sumatra A Large Island*). But the most likely country is Ceylon (Sri Lanka): on the sixteenth-century editions of Pliny's maps the likeness is noticeable; and there is also the fact that it once possessed an ancient port with the vaguely homophonous name 'Tamraparni'.

TERRA AUSTRALIS

*Also known as Terra Australis Ignota, Terra Australis Incognita,
Terra Australis Nondum Cognita, Brasiliae Australis, Magallanica,
Magellanica, La Australia del Espíritu Santo*

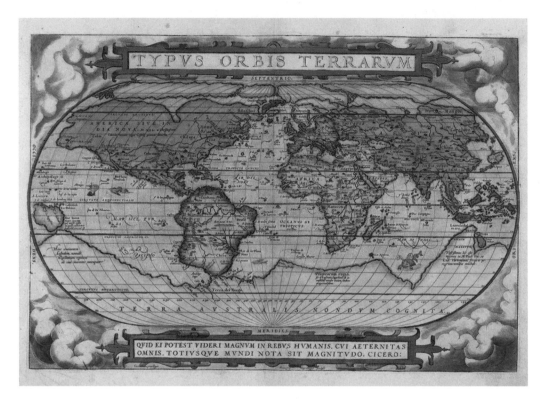

From late antiquity, there was a theory, mentioned
briefly in the Java La Grande entry on page 134, that
the southern hemisphere contained a giant continent
known as the Terra Australis Incognita, that existed as a
counterweight to the mass of territories in the Northern
Hemisphere. Aristotle is one of the earliest sources to
mention the idea in *Meterology* Book 2, Part 5:

*Now since there must be a region bearing the same relation to the
southern pole as the place we live in bears to our pole, it will clearly
correspond in the ordering of its winds as well as in other things. So
just as we have a north wind here, they must have a corresponding
wind from the antarctic. This wind cannot reach us since our own
north wind is like a land breeze and does not even reach the limits
of the region we live in.*

*Ortelius's world map from
Theatrum Orbis Terrarum
(1570) shows the grand
theoretical extent of the
Terra Australis.*

When, in 1520, Ferdinand Magellan sighted land across from the southernmost point of South America – the Tierra del Fuego – the theory was still popularly held, and his findings were mistakenly interpreted by the Spanish sailor Francisco de Hoces as the first sighting of the Austral continent, when he caught his own glimpse in January 1526. With such little known about the Oceania region, the gaps were filled by the theoretical land mass. It was thought the continent extended as far as New Guinea – when the Spanish explorer of the Pacific Álvaro de Saavedra recorded the island and its fellows in 1528, he at first mistook them for the Austral land.

The German polymath Johannes Schöner then produced the first depiction of Terra Australis on his lost globe of 1523 (this assumption is based on the belief that Oronce Finé took much of his information from this globe to produce his double cordiform, or heart-shaped, map of 1531). In 1533, Schöner described the continent, which he calls 'Brasilia Australis', in his *Opusculum geographicum*:

Brasilia Australis is an immense region toward Antarcticum, newly discovered but not yet fully surveyed, which extends as far as Melacha [The Malaysian city of Malacca] and somewhat beyond. The inhabitants of this region lead good, honest lives and are not Anthropophagi [the previously mentioned cannibal race] like other barbarian nations; they have no letters, nor do they have kings, but they venerate their elders and offer them obedience; they give the name Thomas to their children [after St Thomas the Apostle]; close to this region lies the great island of Zanzibar at 102.00 degrees and 27.30 degrees South.

Map of the Pacific Ocean by Ortelius (1589), showing the vastness of the theoretical Terra Australis.

Cornelius Wytfliet, a Flemish geographer, then wrote of the Terra Australis in his 1597 book *Descriptionis Ptolemaicae Augmentum*:

The terra Australis is therefore the southernmost of all other lands, directly beneath the antarctic circle; extending beyond the tropic of Capricorn to the West, it ends almost at the equator itself, and separated by a narrow strait lies on the East

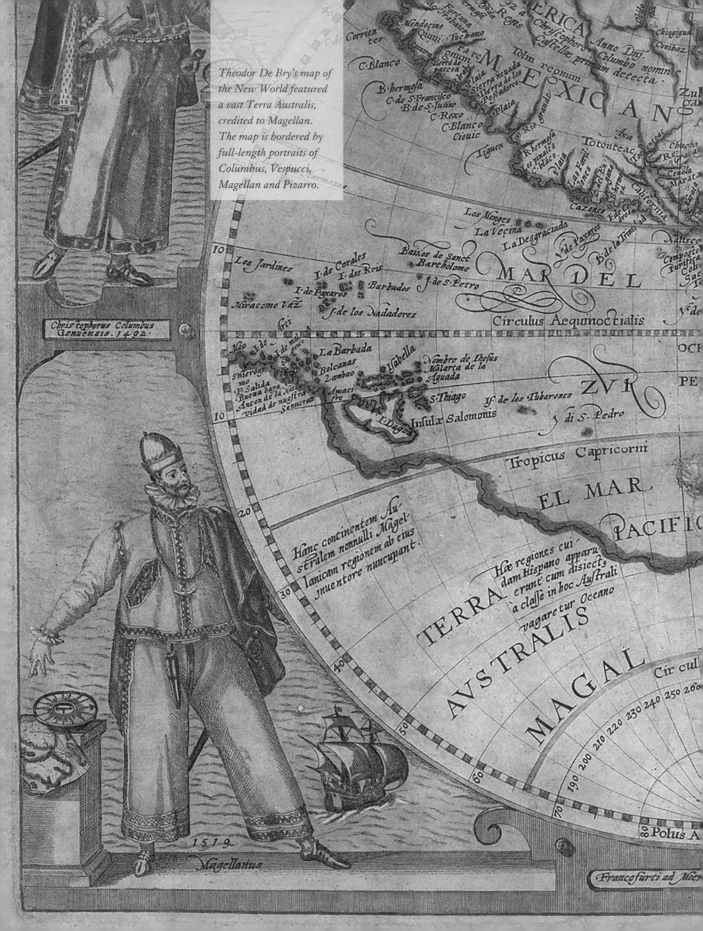

Theodor De Bry's map of the New World featured a vast Terra Australis, credited to Magellan. The map is bordered by full-length portraits of Columbus, Vespucci, Magellan and Pizarro.

Americus Vesputius
Florentinus. 1497.

1526.
Franciscus Pisard.

is Theod. de Bry

The world map featuring Terra Australis that accompanied Joseph Hall's satire Mundus alter et idem sive Terra Australis *(1605).*

opposite to New Guinea ... The terra Australis begins at two or three degrees below the equator and it is said by some to be of such magnitude that if at any time it is fully discovered they think it will be the fifth part of the world.

Ortelius provides the most awesome of depictions on a map of 1589 showing the giant Terra Australis taking up much of the southern regions. The South Pacific is taken up by a vignette of the *Victoria*, Magellan's ship, with his route through the Magellan Strait depicted, with Terra del Fuego incorporated into the southern continent. Cartographic representations of Terra Australis varied wildly in the seventeenth century, and gradually the territory was whittled away as exploration progressed.

Several authors incorporated this popular fascination into their utopian works, including Restif de la Bretonne in *Discoveries in the Southern Hemisphere*. Denis Vairasse's fake travelogue of a voyage to the continent was so persuasive that it fooled many as real, including the reviewer of the prestigious French paper *Journal des Scavants*. It was the perfect setting for dystopia, too. The English bishop and satirist Joseph Hall wrote *Mundus alter et idem sive Terra Australis* in 1605, a Juvenalian satire of contemporary London in which the crew of the *Fantasia* led by Mercurius Britannicus visit the lands of Crapulia (populated by gluttons), Viraginia (nags), Moronia (fools) and Lavernia (thieves).

New Australasian discoveries were at first considered possible evidence of the continent, as was the case with Abel

HOLLANDIA

TERRA AUSTRALIS
Discovered A.D. 1644

TROPIC OF CAPRICORN

NOVA

Discovered 1644

ZEE
LAN
D.A

NOVA

Tasman's sighting of New Zealand in 1642. From here, the story of Terra Australis becomes about the discovery of modern Australia.

Above is the first printed English map of Australia made by Emanuel Bowen, showing the extent of discovery by 1767. Oblivious of Antarctica at this point, Bowen writes at the bottom: 'It is impossible to conceive a Country that promises fairer from its Scituation than this of Terra Australis; no longer incognita, as this Map demonstrates'. Tasman's other great discoveries of 1642 are drawn, including Van Diemens Land (Tasmania) and 'Nova Zeelandia'; and above it his coastal exploration of northern Nova Hollandia (Australia).

In 1814, Matthew Flinders published *A Voyage to Terra Australis*, in which he concludes that the giant southern continent of Aristotle could not exist. Ruling out the myth was important to support his case for transferring the label to the land of New Holland: 'There is no probability', he writes, 'that any other detached body of land, of nearly equal extent, will ever be found in a more southern latitude; the name Terra Australis will, therefore, remain descriptive of the geographical importance of this country, and of its situation on the globe ...'

And so Australia was given its name by the British and the mythical Terra Australis was brought to an end ... or so it was assumed. Just when the idea was finally written off, in 1820, the frozen shores of Antarctica were first sighted beyond the southern floating ice fields, causing a red-faced reevaluation of the dismissed story, and introducing a new focus of obsessive exploration for the years to come.

Emanuel Bowen's A Complete Map of the Southern Continent survey'd by Capt Abel Tafman & depicted by order of the East India Company in Holland *(1744).*

THULE

60°31'N, 28°59'W

*Also known as Thila, Thile, Thoulē, Thula, Thulé,
Tila, Tile, Tilla, Tyle, Tylen, Ultima Thule*

In the early days of exploration the name 'Thule'
represented the vast shadowy unknown of the frozen Far
North – so few details were known about it, yet there
was certainty of its existence. Thule became the label
for every fantasy of what lay at the top of the world, one
of several major, unmarked sections of early charts that
relied upon imagination in place of expeditionary data.

The legend goes back to the writings of the Greek explorer
Pytheas of Massalia, who made a great voyage to explore
northwestern Europe *c.*325 BC, which saw him sail to Britain,
including northern Scotland, and further in a northerly
direction for six days across unknown waters, until finally
he sighted a land he called 'Thule, near the *pepēguia thalatta*'
(solidified sea). (This also came to be marked on maps as the
Mare Congelatii.) Here Pytheas landed and met with the

Sebastian Münster's Tabula
Britanniae *(1571) shows the
island of 'Tihyle or Thule' to the
northeast of the British Isles.*

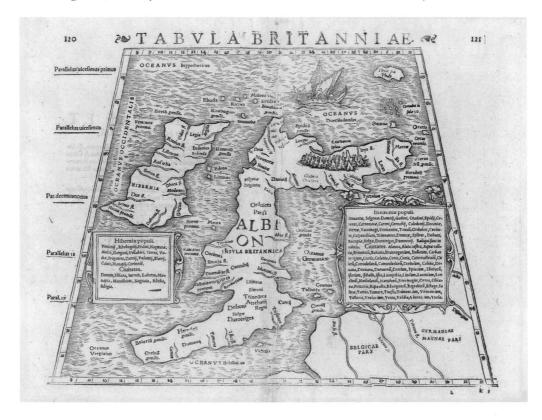

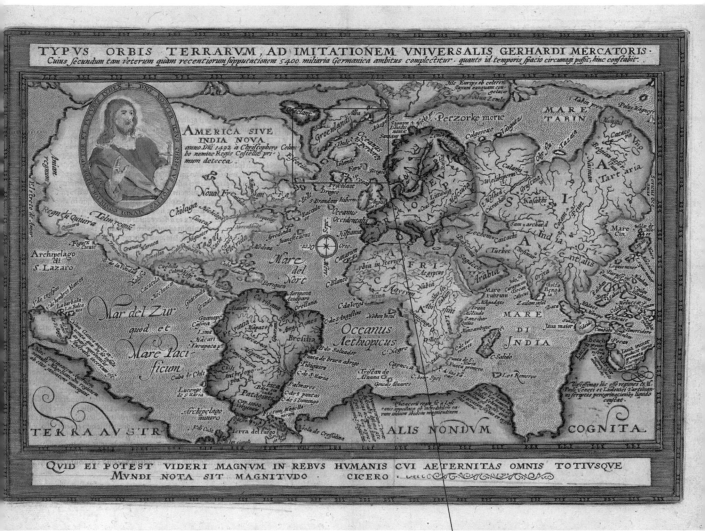

TYPVS ORBIS TERRARVM, AD IMITATIONEM VNIVERSALIS GERHARDI MERCATORIS.
Cuius secundum tam veterum quàm recentiorum supputationem 5400. miliaria Germanica ambitus complectitur. quanto id temporis spacio circumagi possit, hinc constabit.

QVID EI POTEST VIDERI MAGNVM IN REBVS HVMANIS CVI AETERNITAS OMNIS TOTIVSQVE
MVNDI NOTA SIT MAGNITVDO CICERO.

inhabitants, and witnessed their sun setting on the shortest day
and how the land was plunged into darkness during winter.
The land of Thule fired the imaginations of writers: Virgil
refers to it as 'Ultima Thule' (Farthest Thule), a term that
was applied in medieval geography to any distant, unexplored
land. Pliny the Elder describes Thule as 'the most remote of
all those lands recorded', painting it as a place where 'there are
no nights at midsummer when the sun is passing through the
sign of the Crab, and on the other hand no days at midwinter,
indeed some writers think this is the case for periods of six
months at a time without a break.'

Pytheas's account, *About the Ocean* (now lost), had its
doubters, however; Strabo describes Pytheas as an 'arch
falsifier', and writes in Book II, Chapter 5 of his *Geographica*:

*A striking world map by Matthias
Quad from 1600, showing a large
Thule north of Britain, as well as
other geographic misbeliefs such
as Drogeo and Frieslanda (see
Phantom Lands of the* Zeno
Map *entry on page 240).*

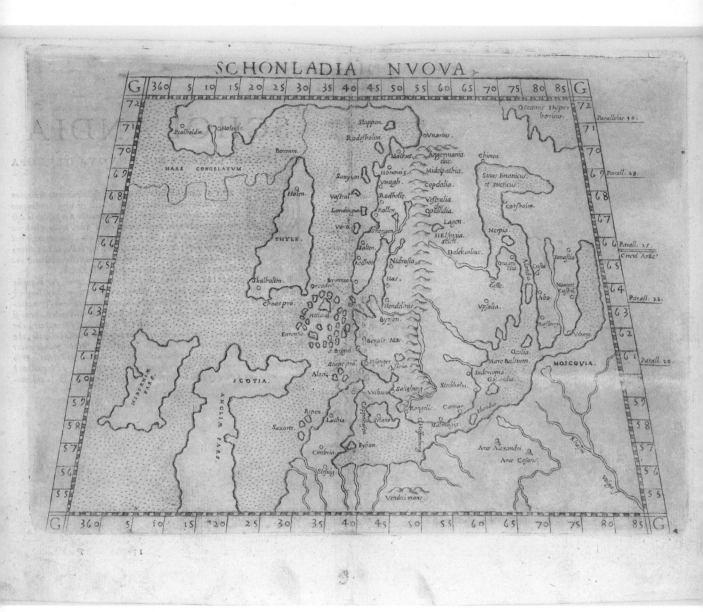

Now Pytheas of Massilia tells us that Thule, the most northerly of the Britannic Islands, is farthest north, and that there the circle of the summer tropic is the same as the Arctic Circle. But from the other writers I learn nothing on the subject — neither that there exists a certain island by the name of Thule, nor whether the northern regions are inhabitable up to the point where the summer tropic becomes the Arctic Circle.

Of the natives of Thule encountered originally by Pytheas, details are preserved by Strabo, who quotes the lost account when writing:

Rusceli's Schonladia Nuova *(1561), one of the earliest printed maps of Scandinavia, shows Thule as a giant island north of Scotland.*

... the people live on millet and other herbs, and on fruits and roots; and where there are grain and honey, the people get their beverage, also, from them. As for the grain, he says, since they have no pure sunshine, they pound it out in large storehouses, after first gathering in the ears thither; for the threshing floors become useless because of this lack of sunshine and because of the rains.

The question of Thule's location would puzzle geographers for centuries. Pomponius Mela placed Thule north of Scythia (a classical name for a region incorporating part of eastern Europe and central Asia); Ptolemy puts it at the Norwegian island of Smøla; and, according to Procopius, Thule was a great island in the north, home to twenty-five tribes, thought to be Scandinavia. Iceland, Greenland, Svalbard and northern Scottish islands have all at one time or another been identified as the original Thule; yet there remains no consensus.

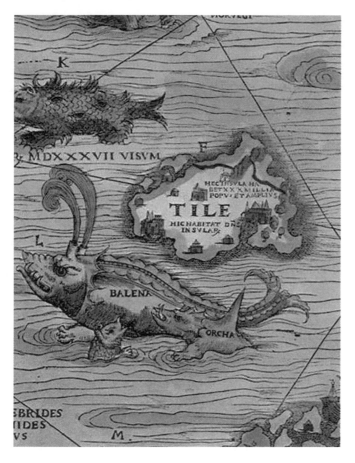

A depiction of Thule *from the* Carta Marina *by Olaus Magnus, which locates the land to the northwest of the Orkney Islands.*

SUNKEN CITY OF VINETA

54°06'N, 14°08'E

Also known as Wineta, Veneta, Weltaba, Jumne, Vimne, Atlantis of the North

The fabled city of Vineta was a bustling, affluent trading post said to lie somewhere in the South Baltic Sea just off the northern German and Polish coast, that in some catastrophe was drowned by waves sometime before 1000 AD and wiped away without a trace. Referred to as 'the Atlantis of the North', the iterations of its mythology vary but all agree that it was an enormous centre of commerce. Bishop Adam of Bremen, the eleventh-century German chronicler, writes of Vineta (which he refers to variously as 'Vimne' and 'Jumne') that it was the largest of all towns in Europe. Bremen records it as a city open for business to all, where Greeks, Slavs, Saxons and barbarians lived together, where the residents were honourable and chaste, hospitable and courteous to strangers (including, at one time, Harald Bluetooth). It was a city rich with the goods of all the nations of the North, with rarities such as the 'volcano pot', otherwise known as the inextinguishable 'Greek fire' incendiary weapon, the composition of which remains unknown. There are also versions of the story told with more of an Old Testament tone, which deride the place as a haven of impenitent sin, claiming this as the reason for its destruction by God, or pirates, or invaders.

On Ortelius's map of the Baltic region, Vineta is marked in the centre.

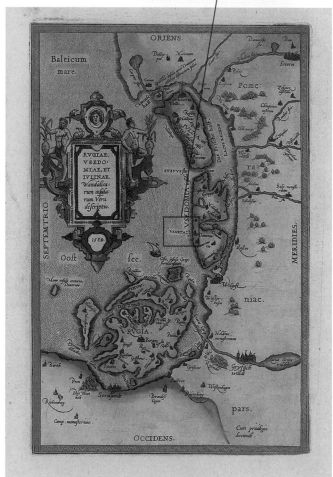

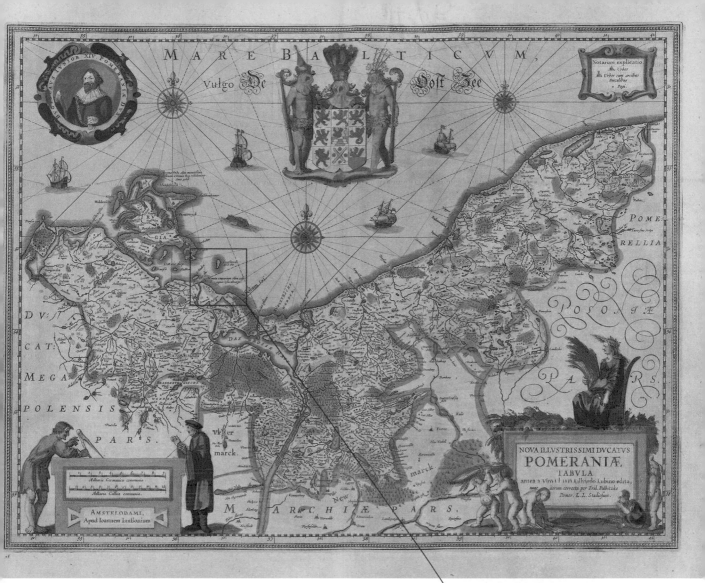

In terms of its geography, the Arab writer Ibrahim Ibn Yaqub (c.970) describes the place – which he calls 'Weltaba' – as 'a large city by the ocean with twelve gates, the greatest of all cities in Europe, farthest northwest in the country of Misiko [Poland] in the marshes by the ocean'. In the sixteenth century, Vineta began to appear on printed maps of the southern Baltic coast. The map shown on page 234 is Ortelius's *Map of Rügen, Usedom and Julin, a true description of the islands of the Vandals,* which places Vineta just north of Germany in the Baltic Sea; one also finds it on Eilhard Lubinus's 1618 *Nova Illustrissimi Principatus Pomeraniæ Descriptio* (map of Pomerania), with a label identifying it as having been destroyed by Conrad, king of Denmark.

On this map of Pomerania from 1640, Hondius marks 'Wineta' on the coast just south of the sea creature.

As one usually finds with remnants of oral tradition, specific details that would help identify its true location are frustratingly scarce. It can't be said for certain if it ever existed, but that hasn't prevented the islands of Wolin and Usedom, as well as the town of Barth, from claiming to be the setting for the story. There is certainly evidence of settlement from the fifth century at Wolin, just off the Polish coast, and it is thought that this is the 'Weltaba' mentioned by Ibn Yaqub. Also, there are the three *vitae* of Otto of Bamberg, which were written in 1140–59, that use the name 'Julin' (similar to Bremen's Jumne) to refer to the site where Wolin would later

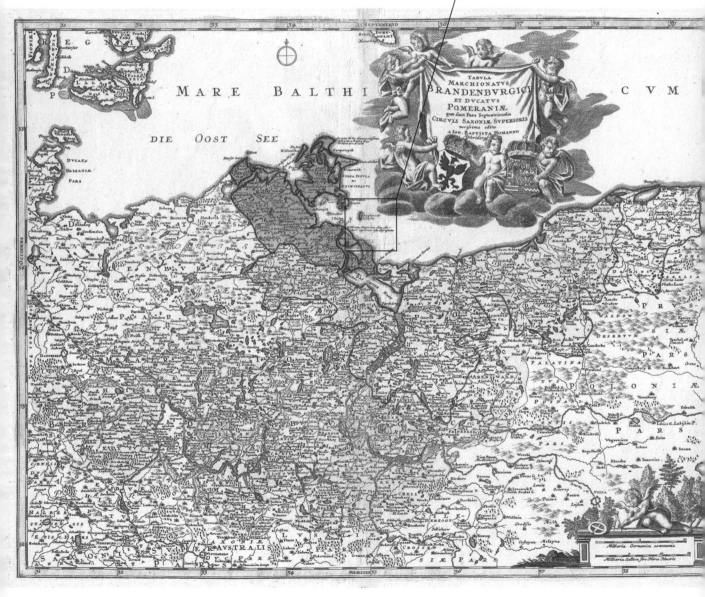

stand. Archaeological evidence, though, is inconclusive, and there is no indication of there ever having been a harbour large enough to accommodate 360 warships, which is another detail mentioned by Bishop Adam of Bremen.

Though proof may be scarce, the legend is cherished by the region's inhabitants. When the seas are still, it is said one can catch sight of Vineta alive beneath the water and glimpse its citizens as they go about their business under the waves. To achieve this perspective, though, one is required to follow a traditional and specific ritual that was relayed by a *New York Times* travel writer from the Baltic Sea island Usedom in August 1897:

A man and a woman fast by day for a week before Easter. Near dawn of Easter morning they take a black cock and a swan, or, if no swan can be got, a perfectly white goose, and proceed in silence to the foot of the Langeberg [mountain]. There they divest themselves of all clothing and march up the sandy cliff to the summit. As the sun rises the man kills the cock and calls on Rhadegast, the woman slays the goose and cries to Svantevit. Then, and not otherwise, do they see the whole of Veneta lying at their feet, as she lay when those gods were worshipped within her gates – Rhadegast, the patron of warriors and pirates; Svantevit, the god of music, prophecy, and women's love.

Opposite: On J. B. Homann's c.1720 map of Brandenburg and Pomerania, 'Wineta' is described in Latin as the famous emporium consumed by waves.

WAK-WAK

38°45'N, 133°19'E

Also known as Island of Waq-Waq, Wák Wák

Mythical islands are as abundant in the histories of Eastern cultures as those of the West. P'eng-lai, for example, which is also known in Japanese mythology as Hōrai, was an island believed by the Chinese to exist somewhere in the eastern end of the Bohai Sea, clustered with four others: Fāngzhàng, Yíngzhōu, Dàiyú and Yuánjiāo. These islands, referred to as the Isles of the Blessed, were a legendary paradise, manifestations of Taoist theories of immortality that were said to hold the elixir of life. Stories were told of native flora with magical properties, able to bestow eternal youth and resurrect the dead. The islands were intensely revered and expeditions to find the Isles of the Blessed are known to have been launched, such as that ordered by Emperor Ch'in Shih Huang-ti in 219 BC.

The story of the Island of Wak-Wak (or Waq-Waq) can be found in some form in the traditions of Turkey, Arabia and India; with references to the island also made in various ninth-century Persian geographical writings that place it somewhere to the east of Korea and China. It is, perhaps, most famously written about in the collection of *One Thousand and One Nights*, in the story of Hassan of Bassorah. After their marriage and the birth of their children, Hassan's wife flees to the island of Wak-Wak, ruled over by her father, the king and commander of an army of 25,000 women, a fearsome group of 'smiters with swords and lungers with lances'. The key features of the island are Mount Wak and the tree that stands at its summit, which grows fruit in the shape of screaming human heads. 'When the sun riseth on them', writes the author of the *Arabian Nights* story, 'the heads cry out all, saying in their cries: "Wak! Wak! Glory be to the Creating King, al-Khallák!" And when we hear their crying, we know that the sun is risen. In like manner, at sundown, the heads set up the same cry, "Wak! Wak! Glory to al-Khallák!" and so we know that the sun hath set.'

In the Arabic work *Aja'ib al-Hind* (*The Book of the Marvels of India*) of *c*.1000, its author Buzurg ibn Shahriyar describes the tree: 'Muhammed ibn Babishad told me that he had learnt

*The mythical Wak-Wak Tree and its human-shaped fruit (*Ta'rikh al-Hind al-Gharbi*, 1729).*

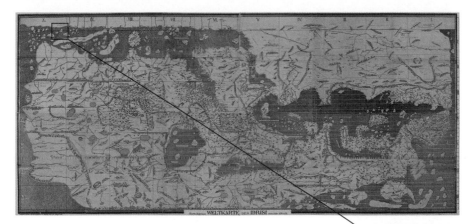

from men who had landed in the country of Waq-Waq, that there is found a species of large tree, the leaves of which are round but sometimes oblong, which bear a fruit similar to a gourd, but larger and having the appearance of a human figure. When the wind shook it there came from it a voice …'

Wak-Wak is drawn as several islands on the map shown above by one of the most respected cartographers of the twelfth century – Muhammad al-Idrisi (1099–1165). He dismisses the fantastical elements ('… there is a tree about which Mas'udi [an Arab geographer] tells us unbelievable stories which are not worth telling'), but describes a land of people living off fish, shell and tortoises, with no gold or ships. The women were naked, save for ivory combs decorated with pearls.

With such exoticism, it's no wonder the story spread so far, to Italian attention: Friar Odorico of Pordenone left Italy to conduct a grand tour of the East in the fourteenth century, and among his journals he mentions hearing rumours of the Wak-Wak tree:

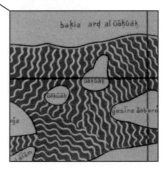

Al-Idirisi drew the Wak-Wak Islands on the Tabula Rogeriana *(1154).*

And here I heard tell that there be trees which bear men and women like fruit upon them. They are about a cubit in measurement, and are fixed in the tree up to the navel, and there they be; and when the wind blows they be fresh, but when it does not blow they are all dried up. This I saw not in sooth, but I heard it told by people who had seen it.

Though the question has never been resolved as to whether Wak-Wak is purely fictitious or a mystery rooted in reality, there have been a variety of theories. Borneo has been suggested as a likely candidate, as have the Sunda Islands, Sumatra, Madagascar, New Guinea and even Australia; while others consider Japan to be the most likely culprit, which seems agreeable when one learns that 'Wo-Kwok' is an old Cantonese term for that country.

PHANTOM LANDS
OF THE *ZENO MAP*

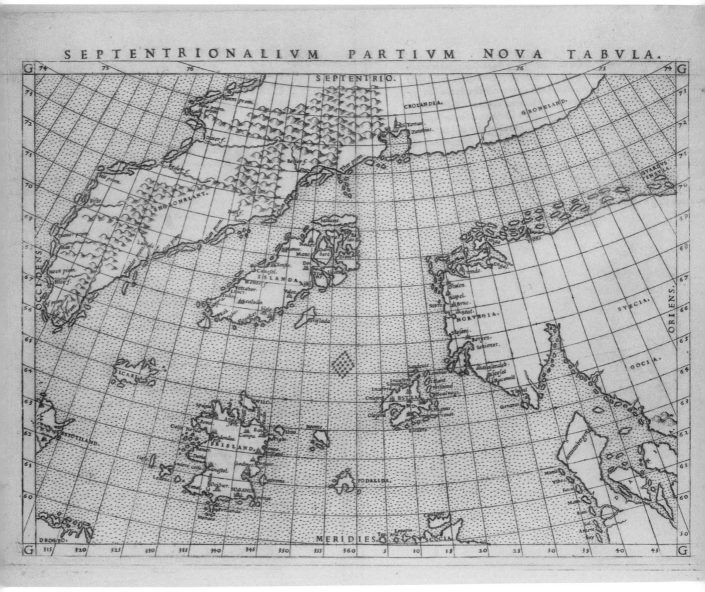

The legendary map of the northern regions, as described in the account of the Zeno brothers (1561).

Nicolò (*c.*1326–*c.*1402) and Antonio Zeno (died *c.*1403) were two Venetian brothers and navigators famous for their pioneering voyage to the North Atlantic, and their subsequent findings of numerous new islands and populations. The Zeno geography had a major impact on later cartography, forming the basis for maps by Ortelius, Mercator and many others. And yet the story is, today, the subject of enormous controversy, because it has been argued, most convincingly, that the Zeno brothers' grand voyage of discovery never happened.

The only source for the adventures of the brothers Zeno (sometimes Zen) is a book written by a sixteenth-century descendant, also called Nicolò Zeno, published by Francesco Marcolini in 1558 and entitled *Dello Scoprimento dell'isole Frislanda, Eslanda, Engrouelanda, Estotilanda, & Icaria, fatto sotto il Polo Artico, da due Fratelli Zeni* (On the Discovery of the Island of Frislanda, Eslanda, Engroenland (sometimes Engroneland), Estotiland & Icaria, made by two Zen Brothers under the Arctic Pole). The contents are letters written by the brothers, the first of which is by Nicolò to Antonio, the second from Antonio to their brother Carlo. Nicolò recounts how he set out in 1380 on an expedition to England and Flanders. His ship was caught in a storm, he and his men were blown off course and came to be wrecked on the shore of Frislanda, an island in the North Atlantic described as larger than Ireland. Here, Nicolò encountered the king of Frislanda, Zichmni, who also ruled over the surrounding islands of Porlanda and Sorant. Nicolò tells of how he advised the Frislandan monarch on his campaign to invade other neighbouring islands, and invites Antonio to join him in Frislanda. Antonio obliges, and joins Nicolò for what would be a fourteen-year martial campaign serving Zichmni.

Antonio headed south of Frislanda and invaded Eslanda, while Zichmni headed north and attempted to land at Islanda, but was beaten back and instead turned his attention to smaller satellite islands off its east coast, claiming seven of them: Bres, Broas, Damberc, Iscant, Mimant, Talas and Trans. Zichmni built a fort on Bres, which he left in the hands of Nicolò, who later sailed to Greenland where he was puzzled to find a monastery with central heating. After four years, Nicolò returned to Frislanda and died.

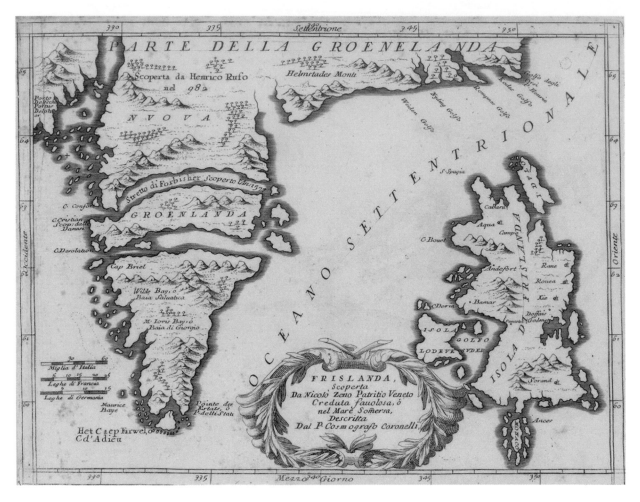

A map of 'Frisland' and
surrounding islands, by
Vincenzo Maria Coronelli
(c.1690), based on the
Zeno discoveries.

Antonio now picks up the narrative, and writes of
encountering in Frislanda a group of fishermen newly
returned from a twenty-five-year expedition. They describe
how, in the far west, they made landfall on a civilized, Latin-
speaking country called Estotiland; and how they discovered a
vastly different island named Drogeo, inhabited by cannibals
and strange animals – this place they managed to escape by
teaching the man-eaters how to fish. On the orders of Zichmni,
Antonio led a force westward and discovered another island,
Icaria. The Icarians refused to allow the Frislandans to make
landfall, threatening to defend their shore to the death, and
so Antonio sailed on, eventually landing at a promontory
called Trin at the southern tip of Engroneland. The sailors
found it inhospitable, but Zichmni was fascinated and decided
to explore further inland. Antonio and his men returned to
Frislanda, leaving Zichmni behind, and Antonio eventually
made his way back to Venice where he died c.1403.

The validity of these accounts provided by Nicolò the younger was initially met with little scepticism. The travels of the Zeno brothers were incorporated in Ramusio's *Delle Navigationi* of 1583; and one also finds them translated into English for Richard Hakluyt's *Divers Voyages* (1582) and the third volume of *Principal Navigations* (1600). Samuel Purchas included an abridged version of the story in *Pilgrimes* (1625), though expressed his doubts as to its authenticity. It wasn't until the nineteenth century that the Zeno legacy and the 'corroborative' claim by Johann Foster in 1784 that Zichmni was, in fact, the navigator Henry Sinclair were challenged. This was done most destructively by Frederick W. Lucas in *The Annals of the Voyages of the Brothers Nicolo and Antonio Zeno in the North Atlantic* (1898), in which he accuses Nicolò the younger of 'a contemptible literary fraud – one of the most successful and obnoxious on record'. Many of the islands mentioned, claims Lucas, were taken from earlier maps such as that by Matheus Prunes and were then thrown around the North Atlantic. As one example, he points to 'Fixlanda' on the Prunes portolan as the inspiration for Frislanda, and suggests both were, in fact, Iceland.

There is also the fact that, at the time Nicolò (the elder) is supposed to have been engaged on these travels, there are records showing that he was in public service in Venice; that he drafted a will in 1400; and that he died *c.*1402. There are also Venetian court papers from 1394 indicating he was on trial for embezzlement committed while military governor of Modone and Corone in Greece between 1390 and 1392.

Frislanda, though possibly being Iceland, was drawn separately on maps (see Gerardus Mercator's map of the Arctic *c.*1570 in the Rupes Nigra entry on page 200). Island has also been suggested as being Iceland. The seven islands that Zichmni annexed – Bres, Broas, Damberc, Iscant, Mimant, Talas and Trans – do not exist. Icaria does not exist. Neither does Esland, although it is perhaps confused with the Shetland Isles. As with so many of the accounts of voyagers studied in this book, the challenge to unpick the truth from a tightly woven tapestry of lies and errors hardened by centuries has proved most difficult in the case of the Zeno brothers and their Frislanda adventure. The debate is ongoing; their contributions to the history of exploration disputed still.

Overleaf: Pieter Goos's magnificent double hemisphere world map from his Sea Atlas of the Water World, *which was published after his death by his wife in 1672.*

uijt gegeven
tot AMSTELDAM bij
Pieter Goos.

SELECT BIBLIOGRAPHY

Adams, P. G. (1980) *Travelers and Travel Liars 1660–1800*, New York: Dover

Babcock, W. (1922) *Legendary Islands of the Atlantic*, New York: Plainview

Beatson, A. (1816) *Tracts Relative to the Island of St Helena: Written During a Residence of Five Years*, London: Printed by W. Bulmer and Co.

Burgh, J. (1764) *An Account of the First Settlement, Laws, Form of Government, and Police, of the Cessares, A People of South America: In Nine Letters, from Mr Vander Neck, One of the Senators of that Nation, to His Friend in Holland*, London: J. Payne

Cameron, I. (1980) *To the Farthest Ends of the Earth: 150 Years of World Exploration*, London: Macdonald

Cherici, P. & Washburn, B. (2001) *The Dishonorable Dr Cook*, Seattle: The Mountaineers Books

Colón, H. (1571) *Historia del Almirante*, Venice

Dalrymple, A. (1775) *A Collection of Voyages Chiefly in the Southern Atlantick Ocean*, London: J. Nourse

Dampier, W. (1697) *A New Voyage Round the World*, London: James Knapton

De Robilant, A. (2011) *Venetian Navigators*, London: Faber & Faber

Delumeau, J. (1995) *History of Paradise: The Garden of Eden in Myth and Tradition*, New York: Continuum Publishing Co.

Dwight, N. (1817) *A System of Universal Geography, for Common Schools: In Which Europe is Divided According to the Late Act of the Congress At Vienna ...*, Albany: Websters & Skinners

Eco, U. (2013) *The Book of Legendary Lands*, London: MacLehose Press

Farini, G. (1886) *Through the Kalahari Desert: A Narrative of a Journey with Gun, Camera, and Note-Book to Lake N'Gami and Back*, London: Sampson, Low, Marston, Searle & Rivington

Flinders, M. (1814) *A Voyage to Terra Australis*, London: G. and W. Nicol

Garfield, S. (2012) *On the MAP: Why the World Looks the Way it Does*, London: Profile

Gould, R. T. (1928) *Oddities: A Book of Unexplained Facts*, London: Philip Allan & Co. Ltd

Hakluyt, R. (1582) *Divers Voyages*, London: Thomas Dawson

Hakluyt, R. (1589) *The Principall Navigations, Voiage and Discoveries of the English Nation, Made by Sea or over Land*, London: G. Bishop & R. Newberie

Horsburgh, J. (1809) *Directions for Sailing to and from the East Indies, China, New Holland, Cape of Good Hope, and the Interjacent Ports*, London: Black, Parry & Kingsbury

Howgego, R. (2003–2013) *Encyclopedia of Exploration*, Sydney: Hordern House

Johnson, D. (1997) *Phantom Islands of the Atlantic*, London: Souvenir

Lucas, F. W. (1898) *The Annals of the Voyages of the Brothers Nicolo and Antonio Zeno in the North Atlantic*, London: H. Stevens Son & Stiles

Mandeville, J. (c.1357) *The Travels of Sir John Mandeville*

Maslen, T. J. (1830) *The Friend of Australia, or, A Plan for Exploring the Interior and for Carrying on a Survey of the Whole Continent of Australia*, London: Hurst Chance

McLeod, J. (2009) *The Atlas of Legendary Lands*, Sydney: Pier 9

Morrell, B. (1832) *A Narrative of Four Voyages ...*, New York: J. & J. Harper

Newton, A. P. (1914) *The Colonising Activities of the English Puritans*, New Haven: Yale University Press

Nigg, J. (1999) *The Book of Fabulous Beasts*, Oxford: Oxford University Press

Nigg, J. (2013) *Sea Monsters: The Lore and Legacy of Olaus Magnus's Marine Map*, Lewes: Ivy Press

Park, M. (1798) *Travels in the Interior Districts of Africa*, London: John Murray

Polke, D. B. (1991) *The Island of California*, Spokane: Arthur H. Clarke

Psalmanazar, G. (1704) *An Historical and Geographical Description of Formosa, an Island Subject to the Emperor of Japan*, London: Wotton, Roper and Lintort

Psalmanazar, G. (1764) *Memoirs of ****, Commonly Known by the Name of George Psalmanazar*, London: R. Davis

Purchas, S. (1625–1626) *Hakluytus Posthumus or Purchas His Pilgrimes, Contayning a History of the World in Sea Voyages and Lande Travells, by Englishmen and others*, London: H. Fetherston

Purdy, John (1822) *Memoir, Descriptive and Explanatory, to Accompany the New Chart of the Ethiopic or Southern Atlantic Ocean, with the Western Coasts of South America, from Cape Horn to Panama: Composed from a Great Variety of Documents, as Enumerated in the Work*, London: R. H. Laurie

Ramsay, R. (1972) *No Longer on the Map*, New York: Viking Press

Ramusio, G. B. (1583) *Delle Navigationi*, Venice: Giunta

Scafi, A. (2013) *Maps of Paradise*, London: British Library

Scott-Elliot, W. (1925) *The Story of Atlantis and the Lost Lemuria*, London: Theosophical Publishing House

Silverberg, R. (1972) *The Realm of Prester John,* Athens: Ohio University Press

Sinclair, D. (2003) *Sir Gregor MacGregor and the Land that Never Was*, London: Headline

Van Duzer, C. (2013) *Sea Monsters on Medieval and Renaissance Maps*, London: British Library

Wafer, L. (1699) *A New Voyage and Description of the Isthmus of America ...*, London: James Knapton

Wellard, J. (1975) *The Search for Lost Worlds*, London: Pan

Williams, G. (2002) *Voyages of Delusion*, London: Harper Collins

Yenne, B. (2011) *Cities of Gold*, Yardley: Westholme

[Zeno, N. & A.] (1558) *Dello Scoprimento dell'isole Frislanda, Eslanda, Engrouelanda, Estotilanda, & Icaria, fatto sotto il Polo Artico, da due Fratelli Zeni*, Venice: Marcolini

INDEX

ACKNOWLEDGEMENTS

I would like to express my deep appreciation to all who provided such indispensable help in the creation of this book: to Charlie Campbell, Ian Marshall, Laura Nickoll and Keith Williams; thank you to Franklin Brooke-Hitching for enduring incessant questions and to my entire family for their support; to Alex Anstey for his artistic contributions and encouragement; to Matt, Gemma and Charlie Troughton, Daisy Laramy-Binks, Kate Awad, Richard Jones and Marie-Eve Poget, Harry Man, Alex Popoff, Katherine Anstey, Alexi Sorrel Harrison, James Miller, Ciara Jameson, Tereza Urbaníková, Luciano Pelizza, Skye Ashton, Georgie Hallett, Thea Lees, Clare Spencer, Andy Murray, June Hogan and Hope Brimelow.

I am especially grateful to those who have been so generous in providing, and allowing the reproduction of, the magnificent works gathered here: Miles Baynton-Williams and Massimo De Martini at Altea Antique Maps Gallery were most generous in allowing the reproduction of many of their images; Barry Ruderman Antique Maps Inc., California, also kindly provided so many crucial images; with further thanks to Maggs Bros Rare Books; John Bonham Rare Books; Richard Fattorini and Francesca Charlton-Jones at Sotheby's; Pom Harrington, Glenn Mitchell and Joe Jameson at Peter Harrington Rare Books; and to Derek McDonnell and Rachel Robarts of Hordern House Rare Books, Sydney.

PICTURE AND MAP CREDITS

Altea Antique Maps Gallery, London
Pg 64, 80, 97, 118, 148, 151, 165, 187, 224, 229, 240

Barry Lawrence Ruderman Antique Maps Inc.
Cover, Pg 12, 15, 16, 20-21, 51, 62, 68, 120, 160, 162, 164 (vignette), 183, 218, 222, 226–227, 230, 242

Boston Public Library
Pg 172

British Library
Pg 58

Hordern House Rare Books, Sydney. Colour by Alex Anstey.
Pg 34 (colour by Alex Anstey), 35, 36 (top and bottom), 37

John Bonham Rare Books
Pg 76, 78

Library of Congress
Pg 43, 66 (Geography and Map Division), 73 (George Grantham Bain Collection), 102 (Geography and Map Division), 124 (Geography and Map Division), 125 (Geography and Map Division), 161 (Geography and Map Division), 219 (Geography and Map Division), 243 vignette (Geography and Map Division)

Maggs Bros. Rare Books
Pg 32, 82

National Library of Australia
Pg 136–7

New York Public Library
Pg 4–5, 14 (Lionel Pincus and Princess Firyal Map Division), 173

Northwestern University Library
Pg 140

Peary-MacMillan Arctic Museum, Bowdoin College
Pg 44

Peter Harrington Rare Books
Pg 166 (bottom)

All other images courtesy of the author.

First published in Great Britain by Simon & Schuster UK Ltd, 2016
A CBS company

Editorial Director: Ian Marshall
Design: Keith Williams, sprout.uk.com
Project Editor: Laura Nickoll

7 9 10 8

Simon & Schuster UK Ltd
1st Floor
222 Gray's Inn Road
London WC1X 8HB

www.simonandschuster.co.uk
www.simonandschuster.com.au
www.simonandschuster.co.in

Simon & Schuster Australia,
Sydney

Simon & Schuster India,
New Delhi

The author and publishers have made all reasonable efforts to contact copyright-holders for permission, and apologise for any omissions or errors in the form of credits given. Corrections may be made to future printings.

A CIP catalogue record for this book is available from the British Library

Hardback ISBN: 978-1-4711-5945-9
Ebook ISBN: 978-1-4711-5947-3

Printed in Italy

MIX
Paper from
responsible sources
FSC® C023419

M. Canaria.

Insulæ Fortunatæ.